Reviews of Environmental Contamination and Toxicology

VOLUME 243

More information about this series at http://www.springer.com/series/398

Reviews of Environmental Contamination and Toxicology

Editor
Pim de Voogt

Volume 243

Springer

DR. PIM DE VOOGT, *Editor*
Reviews of Environmental Contamination and Toxicology

University of Amsterdam
Amsterdam, The Netherlands
E-mail: w.p.devoogt@uva.nl

DR. ERIN R. BENNETT, *Editor*
Bulletin of Environmental Contamination and Toxicology

Great Lakes Institute for Environmental Research
University of Windsor
Windsor, Ontario, Canada
E-mail: ebennett@uwindsor.ca

DR. PETER S. ROSS, *Editor*
Archives of Environmental Contamination and Toxicology

Vancouver Aquarium Marine Science Center
Vancouver, BC, Canada
E-mail: peter.ross@vanaqua.org

ISSN 0179-5953 ISSN 2197-6554 (electronic)
Reviews of Environmental Contamination and Toxicology
ISBN 978-3-319-58723-3 ISBN 978-3-319-58724-0 (eBook)
DOI 10.1007/978-3-319-58724-0

Printed on acid-free paper

This Springer imprint is published by Springer Nature
The registered company is Springer International Publishing AG
The registered company address is: Gewerbestrasse 11, 6330 Cham, Switzerland

Foreword

International concern in scientific, industrial, and governmental communities over traces of xenobiotics in foods and in both abiotic and biotic environments has justified the present triumvirate of specialized publications in this field: comprehensive reviews, rapidly published research papers and progress reports, and archival documentations These three international publications are integrated and scheduled to provide the coherency essential for nonduplicative and current progress in a field as dynamic and complex as environmental contamination and toxicology. This series is reserved exclusively for the diversified literature on "toxic" chemicals in our food, our feeds, our homes, recreational and working surroundings, our domestic animals, our wildlife, and ourselves. Tremendous efforts worldwide have been mobilized to evaluate the nature, presence, magnitude, fate, and toxicology of the chemicals loosed upon the Earth. Among the sequelae of this broad new emphasis is an undeniable need for an articulated set of authoritative publications, where one can find the latest important world literature produced by these emerging areas of science together with documentation of pertinent ancillary legislation.

Research directors and legislative or administrative advisers do not have the time to scan the escalating number of technical publications that may contain articles important to current responsibility. Rather, these individuals need the background provided by detailed reviews and the assurance that the latest information is made available to them, all with minimal literature searching. Similarly, the scientist assigned or attracted to a new problem is required to glean all literature pertinent to the task, to publish new developments or important new experimental details quickly, to inform others of findings that might alter their own efforts, and eventually to publish all his/her supporting data and conclusions for archival purposes.

In the fields of environmental contamination and toxicology, the sum of these concerns and responsibilities is decisively addressed by the uniform, encompassing, and timely publication format of the Springer triumvirate:

Reviews of Environmental Contamination and Toxicology [Vol. 1 through 97 (1962–1986) as Residue Reviews] for detailed review articles concerned with any aspects of chemical contaminants, including pesticides, in the total environment with toxicological considerations and consequences.

Bulletin of Environmental Contamination and Toxicology (Vol. 1 in 1966) for rapid publication of short reports of significant advances and discoveries in the fields of air, soil, water, and food contamination and pollution as well as methodology and other disciplines concerned with the introduction, presence, and effects of toxicants in the total environment.

Archives of Environmental Contamination and Toxicology (Vol. 1 in 1973) for important complete articles emphasizing and describing original experimental or theoretical research work pertaining to the scientific aspects of chemical contaminants in the environment.

The individual editors of these three publications comprise the joint Coordinating Board of Editors with referral within the board of manuscripts submitted to one publication but deemed by major emphasis or length more suitable for one of the others.

Coordinating Board of Editors

Preface

The role of *Reviews* is to publish detailed scientific review articles on all aspects of environmental contamination and associated (eco)toxicological consequences. Such articles facilitate the often complex task of accessing and interpreting cogent scientific data within the confines of one or more closely related research fields.

In the 50+ years since *Reviews of Environmental Contamination and Toxicology* (formerly *Residue Reviews)* was first published, the number, scope, and complexity of environmental pollution incidents have grown unabated. During this entire period, the emphasis has been on publishing articles that address the presence and toxicity of environmental contaminants. New research is published each year on a myriad of environmental pollution issues facing people worldwide. This fact, and the routine discovery and reporting of emerging contaminants and new environmental contamination cases, creates an increasingly important function for *Reviews*. The staggering volume of scientific literature demands remedy by which data can be synthesized and made available to readers in an abridged form. *Reviews* addresses this need and provides detailed reviews worldwide to key scientists and science or policy administrators, whether employed by government, universities, nongovernmental organizations, or the private sector.

There is a panoply of environmental issues and concerns on which many scientists have focused their research in past years. The scope of this list is quite broad, encompassing environmental events globally that affect marine and terrestrial ecosystems; biotic and abiotic environments; impacts on plants, humans, and wildlife; and pollutants, both chemical and radioactive; as well as the ravages of environmental disease in virtually all environmental media (soil, water, air). New or enhanced safety and environmental concerns have emerged in the last decade to be added to incidents covered by the media, studied by scientists, and addressed by governmental and private institutions. Among these are events so striking that they are creating a paradigm shift. Two in particular are at the center of ever increasing media as well as scientific attention: bioterrorism and global warming. Unfortunately, these very worrisome issues are now superimposed on the already extensive list of ongoing environmental challenges.

The ultimate role of publishing scientific environmental research is to enhance understanding of the environment in ways that allow the public to be better informed or, in other words, to enable the public to have access to sufficient information. Because the public gets most of its information on science and technology from internet, TV news, and reports, the role for scientists as interpreters and brokers of scientific information to the public will grow rather than diminish. Environmentalism is an important global political force, resulting in the emergence of multinational consortia to control pollution and the evolution of the environmental ethic. Will the new politics of the twenty-first century involve a consortium of technologists and environmentalists, or a progressive confrontation? These matters are of genuine concern to governmental agencies and legislative bodies around the world.

For those who make the decisions about how our planet is managed, there is an ongoing need for continual surveillance and intelligent controls to avoid endangering the environment, public health, and wildlife. Ensuring safety-in-use of the many chemicals involved in our highly industrialized culture is a dynamic challenge, because the old, established materials are continually being displaced by newly developed molecules more acceptable to federal and state regulatory agencies, public health officials, and environmentalists. New legislation that will deal in an appropriate manner with this challenge is currently in the making or has been implemented recently, such as the REACH legislation in Europe. These regulations demand scientifically sound and documented dossiers on new chemicals.

Reviews publishes synoptic articles designed to treat the presence, fate, and, if possible, the safety of xenobiotics in any segment of the environment. These reviews can be either general or specific, but properly lie in the domains of analytical chemistry and its methodology, biochemistry, human and animal medicine, legislation, pharmacology, physiology, (eco)toxicology, and regulation. Certain affairs in food technology concerned specifically with pesticide and other food-additive problems may also be appropriate.

Because manuscripts are published in the order in which they are received in final form, it may seem that some important aspects have been neglected at times. However, these apparent omissions are recognized, and pertinent manuscripts are likely in preparation or planned. The field is so very large and the interests in it are so varied that the editor and the editorial board earnestly solicit authors and suggestions of underrepresented topics to make this international book series yet more useful and worthwhile.

Justification for the preparation of any review for this book series is that it deals with some aspect of the many real problems arising from the presence of anthropogenic chemicals in our surroundings. Thus, manuscripts may encompass case studies from any country. Additionally, chemical contamination in any manner of air, water, soil, or plant or animal life is within these objectives and their scope.

Manuscripts are often contributed by invitation. However, nominations for new topics or topics in areas that are rapidly advancing are welcome. Preliminary communication with the Editor-in-Chief is recommended before volunteered review manuscripts are submitted. *Reviews* is registered in WebofScience™.

Inclusion in the Science Citation Index serves to encourage scientists in academia to contribute to the series. The impact factor in recent years has increased from 2.5 in 2009 to almost 4 in 2013. The Editor-in-Chief and the Editorial Board strive for a further increase of the journal impact factor by actively inviting authors to submit manuscripts.

Amsterdam, The Netherlands
January 2015

Pim de Voogt

Contents

Aluminium Toxicity to Plants as Influenced by the Properties of the Root Growth Environment Affected by Other Co-Stressors: A Review 1
Joanna Siecińska and Artur Nosalewicz

ARSENIC: A Review on Exposure Pathways, Accumulation, Mobility and Transmission into the Human Food Chain 27
Beste Arslan, Mustafa B.A. Djamgoz, and Ertan Akün

Modelling the Release, Transport and Fate of Engineered Nanoparticles in the Aquatic Environment – A Review 53
Adriaan A. Markus, John R. Parsons, Erwin W.M. Roex, Pim de Voogt, and Remi W.P.M. Laane

Impact of Veterinary Pharmaceuticals on the Agricultural Environment: A Re-inspection 89
Paulina Łukaszewicz, Joanna Maszkowska, Ewa Mulkiewicz, Jolanta Kumirska, Piotr Stepnowski, and Magda Caban

Cadmium Bioaccumulation in Aquatic Oligochaetes Using a Biodynamic Model: A Review of Values of Physiological Parameters and Model Validation Using Laboratory and Field Bioaccumulation Data 149
Leire Méndez-Fernández, Pilar Rodriguez, and Maite Martínez-Madrid

Index 173

Contributors

Ertan Akün Faculty of Engineering, Biotechnology Research Centre, Cyprus International University, Nicosia, Northern Cyprus, Turkey

Beste Arslan Faculty of Engineering, Biotechnology Research Centre, Cyprus International University, Nicosia, Northern Cyprus, Turkey

Magda Caban Faculty of Chemistry, Institute for Environmental and Human Health Protection, University of Gdansk, Gdańsk, Poland

Mustafa B.A. Djamgoz Faculty of Engineering, Biotechnology Research Centre, Cyprus International University, Nicosia, Northern Cyprus, Turkey

Division of Cell and Molecular Biology, Neuroscience Solutions to Cancer Research Group, Imperial College London, London, UK

Jolanta Kumirska Faculty of Chemistry, Institute for Environmental and Human Health Protection, University of Gdansk, Gdańsk, Poland

Paulina Łukaszewicz Faculty of Chemistry, Institute for Environmental and Human Health Protection, University of Gdansk, Gdańsk, Poland

Adriaan A. Markus Deltares, Delft, The Netherlands

Earth Surface Science, IBED, University of Amsterdam, Amsterdam, The Netherlands

Maite Martínez-Madrid Department of Genetics, Physical Anthropology and Animal Physiology, University of the Basque Country, Bilbao, Spain

Joanna Maszkowska Faculty of Chemistry, Institute for Environmental and Human Health Protection, University of Gdansk, Gdańsk, Poland

Leire Méndez-Fernández Department of Zoology and Animal Cell Biology, University of the Basque Country, Bilbao, Spain

Ewa Mulkiewicz Faculty of Chemistry, Institute for Environmental and Human Health Protection, University of Gdansk, Gdańsk, Poland

Artur Nosalewicz Institute of Agrophysics, Polish Academy of Sciences, Lublin, Poland

John R. Parsons Earth Surface Science, IBED, University of Amsterdam, Amsterdam, The Netherlands

Pilar Rodriguez Department of Zoology and Animal Cell Biology, University of the Basque Country, Bilbao, Spain

Erwin W.M. Roex Deltares, Delft, The Netherlands

Joanna Siecińska Institute of Agrophysics, Polish Academy of Sciences, Lublin, Poland

Piotr Stepnowski Faculty of Chemistry, Institute for Environmental and Human Health Protection, University of Gdansk, Gdańsk, Poland

Pim de Voogt Aquatic Environmental Ecology, IBED, University of Amsterdam, Amsterdam, The Netherlands

KWR Watercycle Research Institute, Nieuwegein, The Netherlands

Aluminium Toxicity to Plants as Influenced by the Properties of the Root Growth Environment Affected by Other Co-Stressors: A Review

Joanna Siecińska and Artur Nosalewicz

Contents

1 Introduction 2
2 Mechanisms of Aluminium Toxicity 3
3 Morphological Changes in Roots 3
4 Aluminium and Channel Interactions 4
5 Aluminium-Induced Changes at the Cell Wall 4
6 DNA Damage Under Aluminium 5
7 Plant Mechanisms of Al Tolerance 6
7.1 External Mechanisms 6
7.2 Internal Mechanisms 8
8 Aluminium Toxicity to Plants as Affected by Soil Environment Properties 8
8.1 Aluminium and Drought Interaction 8
8.2 Aluminium and Oxidative Stress 9
8.3 Aluminium Interaction with Heavy Metals 10
8.4 Aluminium Interaction with Soil Nutrients 10
8.5 Aluminium and Nitrogen Interaction 11
8.6 Aluminium and Calcium Interaction 11
8.7 Aluminium and Magnesium Interaction 12
8.8 Aluminium and Phosphorus Interaction 13
8.9 Aluminium and Silicon Interaction 13
8.10 Aluminium and Boron Interaction 16
9 Conclusion 17
10 Summary 17
References 18

J. Siecińska • A. Nosalewicz (✉)
Institute of Agrophysics, Polish Academy of Sciences, Doświadczalna 4, 20-290 Lublin, Poland
e-mail: j.wrobel@ipan.lublin.pl; a.nosalewicz@ipan.lublin.pl

P. de Voogt (ed.), *Reviews of Environmental Contamination and Toxicology Volume 243*, Reviews of Environmental Contamination and Toxicology 243, DOI 10.1007/398_2016_15

1 Introduction

Approximately 30% of the world's total land area consists of acid soils, and it has been evaluated that over 50% of the world's potential arable lands have a pH below 5.0 (Zheng 2010).

Trivalent aluminium (Al^{3+}) is the most abundant metallic element and is ranked as the third most abundant of all the elements in the earth's crust (Bhalerao and Prabhu 2013). Aluminium is primarily incorporated into organic-mineral complexes, mineral surfaces or other non-soluble soil components such as a colloids. Aluminosilicate clays and aluminium hydroxide minerals occurring in alkaline soils are not harmful to plants although there are reports suggesting that aluminate, $Al(OH)_4^-$, can negatively affect plant growth under certain conditions (Kopittke et al. 2005). As the soil pH decreases below 5.0 some of these compounds begin to dissolve releasing soluble mononuclear aluminium cations and other toxic aluminium compounds (Panda and Matsumoto 2007). The toxicity of aluminium depends on soil pH, the chemical structure and concentration of the soluble aluminium compounds in the soil solution and also on the ionic strength of the solution.

Several agronomic strategies have been proposed to manage acid soils. The most important method is the application of lime (calcium carbonate) or similar compounds. Addition of lime to acid soil produces an increase in soil pH which results in lower aluminium toxicity and increased phosphorous availability. Another strategy to improve the productivity of acid soils is the application of organic matter. Organic matter can form aluminium—organic acid complexes, thus reducing aluminium solubility. Wahyudi and Handayanto (2015) reported that legume (*Gliricidia sepium* and *Tithonia diversifolia*) tree prunings can decrease the concentration of plant-available aluminium and improve phosphorous availability in acid soils. However, these soil improvements are not practical in many regions due to relatively high costs and poor transport infrastructure (Abd El-Azeem et al. 2013). Therefore, it is also important to examine the mechanisms of tolerance to aluminium present in some plants, especially in cereals and other important food crops.

In field conditions the plant stress that originates from the toxicity of aluminium in acid soils is often associated with other abiotic stresses. The exposure of plants to many stresses does not usually produce an additive effect, but rather a complex, difficult to predict response that depends on the stress duration, intensity and plant specific response to those stresses. In this review, we highlight recent progress in understanding the effects of aluminium on the morphology and physiology of plants and the resulting changes in response to other environmental stresses e.g. drought, heavy metals, or insufficient nutrition.

2 Mechanisms of Aluminium Toxicity

According to Bian et al. (2013) the effects of aluminium toxicity in plants can be divided into two categories; morphological and physiological. Morphological effects refer to the visual symptoms and damage occurring in different plant tissues and physiological effects refer to changes in metabolism and function and the resultant consequences. Aluminium causes many visible changes in the plant root; it also affects shoot and leaf morphology. However, the main symptom of aluminium stress in plants is the inhibition of root elongation via interaction of aluminium with cells at the root apices. The degree to which aluminium inhibits root growth depends on the plant species, the specific properties of the growth environment and the concentration and form of aluminium ions (Delhaize and Ryan 1995).

3 Morphological Changes in Roots

Inhibition of root growth is the primary symptom of aluminium toxicity; cells in the root apices are very susceptible tissues, they are critical to the onset of aluminium toxicity (Ryan et al. 1993; Meriño-Gergichevich et al. 2010). Root apices accumulate more aluminium than other parts of the roots especially when their growth is inhibited (Kochian 1995). Aluminium stress also causes the roots to become darker, swollen, cracked, and more brittle (Vardar et al. 2006). Long-term exposure to aluminium can affect the architecture of the whole root system, due to the inhibition of primary root elongation and reduced lateral root formation. Roots become stubby due to the inhibition of both cell division and cell elongation (Samac and Tesfaye 2003; Scheffer-Basso and Prior 2015). Clune and Copeland (1999) observed that seedlings of rapeseed (*Brassica napus* L.) exposed to 100 μM of aluminium became stunted and thicker with new lateral roots forming closer to the tap root apex. Swollen, curved and discoloured roots are common symptoms of aluminium stress in rice (*Oryza sativa* L.) seedlings (Alvarez et al. 2012). The development of a brownish colour at root apices partially originates from the accumulation of phenolic compounds produced to neutralise the toxic effects of aluminium (Domingues et al. 2013). Another visible symptom of root damage is the presence of cracks that, according to Yamamoto et al. (2001), are caused by differential cell expansion at different layers of the root epidermis. Disintegration of the cells of the epidermis and the cortical layers in pea (*Pisum sativum* L.) roots at toxic concentrations of aluminium was caused by an increase in the outward pressure of the cells within layers of the cortex (Matsumoto 2000; Kopittke et al. 2008). Aluminium also inhibits elongation of the root hair of white clover (*Trifolium repens* L.) at concentrations higher than 2.5 μM and induces damage to the DNA (Gupta et al. 2013). Many of these damaging effects caused by aluminium occur rapidly.

4 Aluminium and Channel Interactions

Root plasma membrane (PM) is one of the first sites to interact with aluminium. The transport of aluminium ions into the cells is possible through the channels localised in the plasma membrane. Xia et al. (2010) described one aluminium transporter in rice—Nrat1 (Nramp aluminium transporter) which facilitates aluminium uptake via the plasma membrane. This transporter belongs to the Nramp family (natural resistance-associated macrophage protein), which are metal-ion transporters.

Aluminium is known to interfere with the function of many transport proteins and with the uptake of nutrients via direct interaction with the transporters and ion channels. For instance, it affects PM H^+-ATPase—an important protein regulating cytoplasmic pH, which is located preferentially at the cortical and epidermal cells of roots. Other functions of this protein include establishing cellular membrane potential in plant cells (Palmgren and Harper 1999). In acid soils the activity of H^+-ATPase is strongly inhibited by the presence of aluminium ions that affect the formation of the trans-membrane H^+ gradient—essential for secondary transport processes (Gupta et al. 2013). The modulation of this pump activity is essential for plant survival in response to environmental stimuli such as soil acidity (Ahn et al. 2001). Zhang et al. (2015) reported that the activity of PM H^+-ATPase isolated from rice roots was seriously inhibited by strong rhizosphere acidification. However, the expression levels of PM H^+-ATPase isoform 7 were upregulated under H^+ stress as compared to the control, which indicates that increased PM H^+-ATPase activity plays an important role in root tolerance to acidic pH stress. This might facilitate expulsion of excess H^+, initiate alkalization of cytoplasm and partly restore habitual cell activity.

Matsumoto and Yamaya (1986) reported reduced uptake of potassium by pea roots at high concentrations of aluminium in soil. Potassium delivery into the root hair and guard cells by K^+ inward channels is inhibited by aluminium in acid soils (Gassmann and Schroeder 1994). Liu and Luan (2001) showed that aluminium blocks the K^+ inward channels at the cytoplasmic side of the plasma membrane, and concluded that aluminium is able to reduce its permeability and change its activation kinetics. In addition, it was shown that aluminium blocks the activity of many other transporters such as magnesium channels, water channels and the transporters that mediate iron and nitrogen uptake (Jarvis and Hatch 1986; Rengel and Robinson 1989; Chang et al. 1998; Milla et al. 2002).

5 Aluminium-Induced Changes at the Cell Wall

Aluminium contributes to an increase in thickness and rigidity of cell walls via an increase in the concentration of the hemicellulosic polysaccharides. This may be one of the mechanisms by which aluminium decreases the cell-wall extensibility in species such as wheat and rice (Yang et al. 2008). The aluminium-induced decrease in cell wall extensibility may contribute to the rapid inhibition of root elongation by aluminium. Zhu et al. (2012) observed a positive correlation between the level of xyloglucan and the concentration of aluminium in the cell wall of *Arabidopsis* sp. Reduced levels of xyloglucan accompanied by lower levels of aluminium binding in the cell wall contribute to lower cell wall rigidity.

6 DNA Damage Under Aluminium

Clarkson (1965) showed that aluminium can disrupt onion (*Allium cepa* L.) roots at certain concentrations. In support of this finding, there are several studies that indicate that the inhibition of root growth is caused by a decrease in the rate of cell production (Cao et al. 2011; Zhang et al. 2014) and alterations in the nuclei. For instance, aluminium can bind to chromatin and alter its structure and disrupt the ability of DNA to replicate (Silva et al. 2000). Clarkson (1969) reported that root growth and mitosis in onions are inhibited after 48 h of exposure to aluminium, which is related to the possible interruption of the S-period of the mitotic cycle. Matsumoto et al. (1977) reported that aluminium accumulated in nuclei became associated with DNA in Al-treated roots of pea (*Pisum sativum* L.). Furthermore, the template activity of DNA was suppressed by aluminium in vitro, which may cause the inhibition of mitoses of pea root cells. Similar conclusions were reported by Sampson et al. (1965) who found that DNA synthesis was strongly inhibited after 4 h of exposing barley (*Hordeum vulgare* L.) cells to aluminium. This mechanism is controlled by ATR (Ataxia telangiectasia mutated related) a protein kinase, that is universally required for monitoring DNA integrity in higher eukaryotes. It was shown that decreases in AtATR expression can greatly enhance aluminium tolerance allowing the plants to grow normally in soils that contain toxic levels of aluminium (Rounds and Larsen 2008).

It is noteworthy that aluminium can also affect cells in the mitosis cycle and alter the structures of the chromosomes leading to the dysregulation of the cell cycle (Nezames et al. 2012). Zhang et al. (2014) explored the toxic effect of aluminium on the chromosome behaviour of Masson pine cells. They observed four chromosomal aberrations including anaphase bridge formation, C-mitosis, chromosome fragmentation and chromosome stickiness that resulted in pathological cell death and hence the inhibition of cell replication. The main cause of these aberrations appears to be the production of reactive oxygen species (ROS) that can bind to purine and pyrimidine bases in DNA. Nonetheless low, non-toxic concentrations of Al may play a positive role in protection against DNA damage; that process is mediated by ROI (Reactive Oxygen Intermediate) (Achary and Panda 2010). From studies on onion, Achary et al. (2013) concluded that low concentrations of aluminium may protect DNA from damage resulting from exposure to the methylmercury chloride toxin (formerly used in pesticides).

Recent studies on sorghum (*Sorghum vulgare* L.) and maize (*Zea mays* L.) also showed that aluminium induces epigenetic changes to DNA like hypomethylation or hypermethylation of cytosine, which contributes to epiallelic diversity and modulation of gene regulations (Kimatu et al. 2011; Kimatu 2015). Similarly, studies by Dong et al. (2006) and Choi and Sano (2007) demonstrated that aluminium stress increases cytosine methylation of DNA in tobacco (*Nicotiana tabacum* L.) and rice. These modifications of chromatin methylation may have a range of implications such as gene silencing that can affect plant growth and functioning along with the response to stresses.

7 Plant Mechanisms of Al Tolerance

Plants have the ability to adapt to various environmental stresses including aluminium toxicity and they display many physiological mechanisms of adaptation. The mechanisms of aluminium resistance include external (exclusion or apoplastic) strategies, in which aluminium is prevented from entering the root cells by means of physical or biochemical barriers, or internal (symplastic) strategies, which involve the detoxification of aluminium that enters the root cells by chelating the aluminium, storing it in a vacuole or otherwise preventing it from interacting with the sensitive components of the cell (Delhaize and Ryan 1995). Aluminium tolerance of species important for agriculture can be increased by genetic selection and breeding using recurrent selection and intercrossing to transfer traits from more tolerant germplasm to elite cultivars, or it can be increased using genetic engineering by transferring the known aluminium tolerance genes from one species to another. These techniques are based on the screening of plant cell cultures for aluminium-tolerant genotypes of plants, or the cytogenetic approach is adopted (de Camargo and Filho 2001; Ezaki et al. 2001).

7.1 External Mechanisms

External mechanisms of aluminium tolerance are also known as aluminium exclusion mechanisms, because they prevent the uptake of aluminium into the symplast (Simões et al. 2012). These mechanisms include the binding of aluminium to different cell structures such as the epidermis, cortex and root apoplast, increasing the pH of the rhizosphere, exudation of many chelating substances or even interaction with mycorrhizal fungi.

One of the external barriers limiting aluminium toxicity is the root cell wall, which is able to accumulate aluminium due to the presence of negatively charged carboxylic groups in the pectin and hemicellulose (Liu et al. 2008). Although the aluminium affects the structure of cell walls, most aluminium ions are immobilised in them, thereby protecting other structures. The Casparian band in the endodermis contains water-impermeable deposits of suberin that regulate water and mineral uptake by the roots. Silva et al. (2010) examined the differentiation of the endodermis in two wheat cultivars characterised by different sensitivities in response to aluminium addition and observed that aluminium-tolerant wheat had higher endodermis differentiation at the hair region of the roots. This supports the idea that these changes may be involved in controlling aluminium accumulation or partitioning in tolerant genotypes.

Another important mechanism of increasing plant aluminium tolerance involves exudation of organic acid anions such as malate, citrate and oxalate to soil (Sasaki et al. 2004; Delhaize et al. 2004). These organic acid anions are able to bind to aluminium ions thereby saving the roots from damage and maintaining their growth and some reports suggest that may play a role in mineral nutrient acquisition and transport (Ligaba et al. 2012). Detoxification of aluminium is also possible through the formation of very stable rings; five-, six-, and to a lesser extent seven-membered structures between organic acid ions and aluminium (Ma and Ryan 2010). The amount of organic acid ions released depends on the plant species, concentration of aluminium and the duration of exposure (Delhaize and Ryan 1995). Wang et al. (2014) observed citrate exudation in various ecotypes of barrel clover (*Medicago truncatula* L.) and concluded that the higher rate of exudation of citrate by ecotype Jemalong A17 compared to ecotype R108 provided greater tolerance to aluminium. Soybean (*Glycine max* (L.) Merrill) cultivars have also been shown to release organic acid ions under aluminium stress; the increased activity of an enzyme involved in the synthesis of citrate was observed in aluminium-tolerant cultivars treated with aluminium (Zheng et al. 2014). These results indicate that changes in the activity of enzymes that mediate the synthesis of organic acid ions play an important role in aluminium tolerance mechanisms. When enzymes involved in malate and citrate synthesis are over-expressed in transgenic alfalfa (*Medicago sativa* L.) plants, it results in increased organic acid biosynthesis and increased aluminium tolerance compared to non-transgenic plants (Tesfaye et al. 2001). Zhao et al. (2003) observed a rapid (within 2 h) release of organic acid ions after exposing barley to aluminium. Citrate secretion from the root tips of barley significantly increased at low soil temperatures. Despite many studies conducted to date, the exact regulation mechanism of the exudation of organic acid ions is still poorly understood.

Oxalate is another strong detoxifier of aluminium, and it has been shown to maintain root growth in acid soils. A higher oxalate exudation rate was observed in the tolerant varieties of birdsfoot trefoil (*Lotus corniculatus* L.) allowing for the effective chelation of aluminium and prevention of its entry into the root (Pal'ove-Balang et al. 2012).

Root mucilage has also been known to immobilise toxic metal cations within the rhizosphere (Horst et al. 1982). These gelatinous substances consisting mainly of polysaccharides are exuded from the outer layers of the root cap. Detoxification of aluminium by mucilage is possible by inactivation of aluminium with polysaccharides e.g. uronic acids. Watanabe et al. (2008) found that mucilage from banks melastoma (*Melastoma malabathricum* (L.) Smith) can facilitate aluminium uptake and selectively increase nutrient uptake by this species. Cai et al. (2013) concluded that the higher capacity to exclude aluminium in an aluminium-resistant soybean cultivar is related to the immobilisation of aluminium by mucilage secreted from root border cells. Some reports suggest that arbuscular fungi that are obligate symbionts might play a protective role for plants exposed to aluminium toxicity (Cumming and Ning 2003). Similarly Rouphael et al. (2015) showed that arbuscular mycorrhizal fungi are also effective in alleviating aluminium toxicity in pumpkin (*Cucurbita pepo* L.).

7.2 Internal Mechanisms

Plants have also evolved internal mechanisms, which allow them to detoxify aluminium that enters the root cells. The concentration of aluminium in the symplast is low compared to the apoplast but still potentially toxic and damaging. Therefore, the internal mechanisms of detoxification are essential for plant growth and development. These internal mechanisms of aluminium tolerance include aluminium chelation by organic acids in the cytosol and sequestration of aluminium into vacuoles by half-size ABC transporters located at the tonoplast (Huang et al. 2012).

Upon entering the root cells, aluminium is chelated by organic acid anions and in this form is transported from the root to the shoot (Brunner and Sperisen 2013). Apart from organic acids other compounds, such as phenol derivatives can also chelate aluminium in the cytosol (Osawa et al. 2011).

Enhanced vacuolation of root cells during aluminium exposure was observed in many plant species including barley (Lee et al. 2015) and in tobacco cells (Panda et al. 2008). Enhanced vacuolation was also observed in the cells of the peripheral cap, apical meristem and cortex 12 h after treatment with 0.05 mM aluminium, indicating that aluminium treatment may promote the autophagy of cells (Ikeda and Tadano 1993). The internal immobilisation of aluminium appears to be affected by the degree of pectin methylation in the cell wall and it is noteworthy that some aluminium-tolerant plants have a higher degree of pectin methylation compared to aluminium sensitive plants, which provides evidence that pectin methylation reduces the accumulation of aluminium in the cell wall (Brunner and Sperisen 2013).

8 Aluminium Toxicity to Plants as Affected by Soil Environment Properties

8.1 Aluminium and Drought Interaction

The combination of drought and aluminium stress is relatively common in regions where acid soils typically occur and is associated with complicated soil-plant interactions. Both water deficit and aluminium toxicity can affect plant growth and development. The inhibition of root growth by aluminium increases the susceptibility of plants to water deficit (Samac and Tesfaye 2003). Many reports (Wang et al. 2006; Manavalan et al. 2009) presented both, negative and positive plant responses to aluminium when these two stress factors occurred simultaneously. Slugeňová et al. (2011) examined physiological responses of Norway spruce (*Picea abies* (L.) Karst) seedlings to drought and aluminium. They reported that water deficit alone can reduce the net photosynthetic rate as a consequence of stomatal closure. However, they did not find statistically significant differences in photosynthesis values between plants exposed to drought alone and a combined drought and aluminium treatment. Nevertheless, their results indicated that the impact of drought on physiological processes (e.g. photosynthesis, fluorescence) in Norway spruce seedlings was enhanced by the presence of aluminium. Other studies conclude that drought can diminish aluminium toxicity. For instance, Yang et al. (2011) found that the water deficit induced by PEG 6000 inhibited aluminium accumulation in the root tip of the aluminium-sensitive common bean. This improved aluminium tolerance was associated with several cell wall-modifying and cell wall-assembling-related genes such as xyloglucan endotransglycosylase precursors, glucan endo-1,3-b-glucosidase precursors and hydroxyproline-rich glycoprotein which help to limit aluminium uptake into the root apices (Yang et al. 2011). A different trend was reported by Schier and McQuattie (2000) who concluded that a decrease in soil water content increased aluminium concentration in soil and enhanced aluminium toxicity. Goldman et al. (1989) showed that combining water deficit and aluminium stress affects many plant functions, including transpiration rate, to a greater extent than these stresses individually. These contrasting reports show that further studies are required. Dry soils also influence penetration resistance, so the combination of dry soils with aluminium toxicity might have synergistic effects on root growth (Whitmore et al. 2011). Damage occurring within the root tip zone in the presence of aluminium might additionally limit the root growth rate depending on the soil type (Bengough et al. 2011).

The response to a combination of aluminium toxicity and water deficit may vary according to species-dependent water uptake pathways. Apoplastic uptake is strongly affected by transpiration and not as specific to transported chemicals compared to symplastic water uptake pathways (White and Broadley 2003). Thus, the drop in soil water potential at which a specific plant species or cultivar closes its stomata will have an effect on aluminium toxicity.

8.2 *Aluminium and Oxidative Stress*

Oxidative stress is defined as a disturbance of the equilibrium between the oxidant and antioxidant components. A common factor between most stresses (drought, salinity, low pH) is the enhanced production of Reactive Oxygen Species (ROS) and a peroxidation of lipids. The increased concentration of ROS is balanced by antioxidant enzymes, which maintain the oxidative stability in the cells (Singh et al.

2012). Aluminium and oxidative stress are strongly interrelated in plant cells. Aluminium-induced ROS production causes the oxidation of many cell components, e.g. protein, nucleic acids and membrane lipids, which act as the primary indicators of oxidative stress (Cakmak and Horst 1991). It is well documented that aluminium-induced oxidative stress occurs in many plant species e.g. barley (Tamás et al. 2006), maize (Boscolo et al. 2003), pea (Matsumoto and Motoda 2012) or peanut (Huang et al. 2014).

8.3 Aluminium Interaction with Heavy Metals

Acidification of soil affects the transformation and biogeochemical cycling of heavy metals through its effect on the physical, chemical and biological characteristics of soils. Soil acidification may increase the bioavailability of some trace heavy metals by the transformation of heavy metal ions via modification of the surface charge in variable charge soil particles; this can alter the speciation of metals and influence the reduction and oxidation reactions of these metals (Bolan et al. 2003). Guo et al. (2005) showed a significant increase in the concentration and mobility of Cd, Zn and Cu at lower pH in two acidic forest soils from the Hunan province in China. Similar results were reported by Wu and Zhang (2002) for Cd availability in low pH soils. Although the effects of cadmium toxicity on plants are well established, our understanding of the interactions between Cd toxicity and soil acidity is incomplete. This is especially the case for the combined effects of cadmium and aluminium on plants (Guo et al. 2007; Shamsi et al. 2007). Guo et al. (2004) reported an antagonistic interaction between cadmium and aluminium since the aluminium content in the aluminium-sensitive and tolerant genotypes of barley was lower in roots and leaves compared to the control when cadmium was also present. This indicates that Cd reduced the accumulation of aluminium in the shoots compared to the plants treated with aluminium alone.

8.4 Aluminium Interaction with Soil Nutrients

Acidic soils can also limit plant growth and development by inducing nutrient deficiencies, high concentrations of soluble aluminium in acid soils can induce P, Ca and Mg deficiency in shoots and leaves by interfering with the uptake, transport and utilisation of nutrients (Scholl et al. 2005).

8.5 Aluminium and Nitrogen Interaction

Experiments on acid soils indicate that there is a strong relationship between Al and NO_3^- in the soil solution of acidic soils (Gundersen and Rasmussen 1990). The concentration of HNO_3^- in soil can be increased by nitrification leading to further acidification and aluminium release to the soil solution. In addition, changes in the NO_3^- to NH_4^+ ratio resulting from N uptake by plants affect the pH of the rhizosphere and in consequence, the aluminium concentration (Taylor 1988). Some report indicated that effect of Al on NO_3^- uptake and assimilation depends upon the concentration of aluminium. Results of Rufty et al. (1995) at soybean seedling show that the greater aluminium concentration is in solution, the more sever decreased NO_3^- uptake is observed.

8.6 Aluminium and Calcium Interaction

Disturbance of the cell calcium (Ca) homeostasis appears to be an important aspect of ion-related environmental stresses in the presence of soil-available aluminium. Disturbance of cytoplasmic Ca^{2+} homeostasis is believed to be the primary target of aluminium toxicity and may result in the inhibition of cell division or root elongation via disruption of Ca^{2+}-dependent biochemical and physiological processes (Rengel and Zhang 2003). Many reports show an increase in cytosolic calcium under aluminium stress (Zhang and Rengel 1999; Qifu et al. 2002; Bhuja et al. 2004).

Reduction in calcium uptake as a result of aluminium toxicity causes many visible changes similar to calcium deficiency. For instance, a common symptom of aluminium toxicity in shoots is the curling and rolling of young leaves, which is related to calcium deficiency (Foy 1978). Alternatively, Rengel and Zhang (2003) indicated that aluminium-resistant and aluminium-sensitive lines are characterized by an increase in cytosolic calcium in both cultivars studied. Generally, aluminium can affect cytosolic calcium homeostasis but it is still unclear if it is a primary or secondary effect. One of the mechanisms explaining the disruption of calcium homeostasis is associated with calcium channels, which are mainly blocked by aluminium (Huang et al. 1992; Rengel and Elliott 1992). Another explanation of an increase in cytosolic calcium proposed by Rengel and Zhang (2003) is the inhibition of the Ca^{2+}-ATPase pump localized in the endoplasmic reticulum membrane. In acid soils, aluminium also reacts with the cell wall components of the roots leading to the displacement of cytoplasmic calcium by aluminium (Rengel 1996). Blamey (2001) showed that aluminium forms a stronger bond to pectin (carboxyl groups in pectin) than calcium. Calcium acts as the main component in cross-linking the pectic materials in the cell wall so the displacement of pectic-bound calcium by aluminium undoubtedly alters the physiological properties of the cell wall like its extensibility and permeability (Horst 1995; Blamey 2001; Poschenrieder et al. 2008).

Callose is a polysaccharide that allows cementing of the cell walls, thus preventing cell wall loosening. It also prevents the symplastic or apoplastic passage of signalling molecules such as indole-3-acetic acid, which is required for root growth and development (Alvim et al. 2012). Callose deposition in the plant cell walls is another aluminium-induced injury at the cellular level and is strongly connected with the increase in cytosolic Ca^{2+} concentration which is one of the prerequisites for the induction of callose synthesis, but not the only factor modulating the increases in callose synthesis and deposition. The release of calcium ions from subcellular compartments is most likely associated with the activity of free calcium in the cytoplasm, which is the signal for callose formation (Bhuja et al. 2004). Nevertheless, it is not the only factor causing callose formation, since the availability of UDP-glucose and changes to cellulose synthase conformation are also important. Synthesis and accumulation of callose in response to aluminium exposure was well-described by Sivaguru et al. (2000) who showed that production of this polysaccharide blocks plasmodesmata and inhibits the symplastic and apoplastic transport. The accumulation of callose is strongly dependent on the aluminium resistance of the plant. Aluminium-sensitive cultivars accumulate more callose than tolerant ones because they experience more intense stress and perhaps have their membranes damaged to a higher degree (Schmohl and Horst 2000; Too et al. 2014).

8.7 Aluminium and Magnesium Interaction

In acid soils, magnesium deficiency is induced by the leaching of magnesium (Mg) into deep soil layers, below the root zone. The primary interactions between aluminium and magnesium occur in the root apoplast (da Silva et al. 2005). At high external concentrations of aluminium, the major part of the soil magnesium is bound at the binding sites of the root apoplasm (root CEC—Cation Exchange Capacity). As a result, magnesium can be replaced by aluminium, which has negative consequences for magnesium uptake through the apoplastic pathway (Bose et al. 2011). Furthermore, aluminium competes with magnesium for membrane transporters and metal binding sites in many enzymes (Pécsváradi et al. 2009). Pina and Cervantes (1996) demonstrated that under acidic conditions aluminium has a stronger binding affinity to metal binding sites in the ATP than magnesium. Deficiency in magnesium leads to the production of reactive oxygen species (ROS) due to the abnormal functioning of mitochondria (Cakmak and Kirkby 2008) and the reduction of chlorophyll content (Lazarević et al. 2014).

Conversely, high concentrations of magnesium can alleviate aluminium toxicity by increasing the ionic strength of the solution, decreasing aluminium saturation at the apoplastic exchange site, which reduces aluminium activity at the root cell plasma membrane surface and lowers aluminium concentration at the root CEC (Grauer and Horst 1992; Bose et al. 2011).

8.8 Aluminium and Phosphorus Interaction

Phosphorus (P) is one of the most important nutrients for plant growth and metabolism; it is a key component of molecules such us nucleic acids and phospholipids and ATP, taking part in the control over enzymatic reactions and metabolic pathway. The availability of phosphorus in soil strongly depends on the pH of the soil. In acid soils phosphorus forms sparingly-soluble complexes with aluminium and iron (Hinsinger 2001). Aluminium can also bind to phosphorus creating insoluble complexes, especially $Al(PO_4)_3$. Generally, phosphorus can alleviate aluminium toxicity in plants by aluminium-phosphorus precipitation in soil and plants (Liao et al. 2006; Sun et al. 2008). These complexes can accumulate in the root cell walls and retard aluminium uptake into the cytosol (Zheng et al. 2005). In this form aluminium is non-toxic to plants, but the plant suffers from phosphorus deficiency to a greater extent. Guo et al. (2012) showed an inhibited growth of rice seedlings in the presence of toxic concentrations of aluminium and phosphorus deficiency indicating a synergistic effect of P deficiency on aluminium toxicity. However, there are some reports that disagree with this finding (Chen et al. 2012). Tan and Keltjens (1990) found that the addition of phosphorus can alleviate aluminium toxicity only for aluminium-sensitive sorghum cultivars. Zheng et al. (2005) reported that aluminium-resistant cultivars had a greater capability to utilise P due to the active transport of insoluble aluminium-phosphorus deposits into vacuoles, which is claimed to be one of the mechanisms of intracellular detoxification. Studies aimed to evaluate aluminium toxicity at presence of other stressors or environmental factors are summarized in Table 1.

8.9 Aluminium and Silicon Interaction

Silicon, like aluminium, is an abundant element in soils, however in contrast to aluminium, solubility of silicon is unaffected by pH in range 2.0–9.0. Silicon can influence availability of other mineral elements through complex interactions that can be achieved either outside or inside plant cells. There are known three mechanisms of interaction between aluminium and silicon: (a) reduction of aluminium availability, (b) detoxification of internal aluminium and (c) increased pH solution by Si-sources (Pontigo et al. 2015). The reduction of aluminium availability results from formation of HAS (hydroxyaluminosilicate) complexes in the growth media (Hodson and Evans 1995) or in the apoplast of maize (*Zea mays* (L.) Lixis) root apex (Wang et al. 2004). HAS complexes were shown as a helping in

Table 1 Studies on aluminum toxicity including additional interactions with growth environment

Primary stress	Co-stressor or interactions		Effects of interactions	Species	References
Aluminium	Drought		Drought led to a lower Al accumulation in needles Reduced Al accumulation in the root tips Additional inhibition of root elongation at Al + drought	*Picea abies* L. *Phaseolus vulgaris* L.	Slugeňová et al. (2011) Yang et al. (2011) Yang et al. (2011)
	Soil compaction		Negative impact of soil acidity on root growth by aluminium resistant genotypes was greater in compacted soil	*Dactylis glomerata* L.	Haling et al. (2011)
	Heavy metals	Cadmium	Stronger inhibition of seedling growth and increased activity of antioxidant enzymes at presence of Cd + Al than Al Combined application of Cd and Al enhanced their accumulation in roots, but caused a reduction in shoots	*Hordeum vulgare* L. *Glycine max* L.	Guo et al. (2004) Shamsi et al. (2007)
		Manganese	Significantly decrease of A1 accumulation in roots in response to application of Mn	*Triticum aestivum* L.	Blair and Taylor (1997)
		Lead	Inhibition of root elongation by Pb was lessened by the presence of Al	*Picea abies* L.	Godbold and Kettner (1991)
		Iron	Amelioration of Fe toxicity by Aluminium leading to decreased Fe concentration in shoots and roots	*Melastoma malabathricum* L.	Watanabe et al. (2006)
	Plant nutrients	Nitrogen	Aluminium inhibited NO_3^- and NH_4^+ uptake Aluminium reduced nitrate assimilation in roots Al strongly reduced nitrate uptake and also the activity of nitrate reductase in the apical parts of roots NH_4^+-decreased Al accumulation in root tip	*Zea mays* L. *Cucumis sativus* L. *Lotus japonicas* L. *Oryza sativa* L.	Purcino et al. (2003) Jerzykiewicz (2001) Pal'ove-Balang and Mistrik (2007) Wang et al. (2015)
		Calcium	Inhibition of Ca^{2+} transport via symplasm, disruption of Ca^{2+} homeostasis in cytoplasm and displacement of Ca by Al		Meriño-Gergichevich et al. (2010)
		Magnesium	Mg alleviates Al injury in roots Aluminium induced Mg deficiency	*Glycine max* L. *Sorghum bicolor* L.	da Silva et al. (2005) Bose et al. (2011) Tan and Keltjens (1995)

		Phosphorus	P deficiency aggravated Al toxicity to plant growth and induced more severe lipid peroxidation Aluminium induced P deficiency	*Oryza sativa* L. *Sorghum bicolor* L.	Guo et al. (2012) Tan and Keltjens (1990)
		Boron	Boron alleviates Al toxicity in roots Boron supplementation enhances root growth under Al toxicity	*Pisum sativum* L. *Triticum aestivum* L.	Yu and Goldbach (2007) Hossain et al. (2004)
		Silicon	Si reduced shoot and root Al concentration Formation of Al-Si non-toxic complex Si ameliorates inhibition of root elongation	*Solanum tuberosum* L. *Zea mays* L. *Glycine max* L.	Dorneles et al. (2016) Wang et al. (2004) Baylis et al. (1994)
	Other interactions	Oxidative stress	Aluminium induced ROS production	*Pisum sativum* L.	Matsumoto and Motoda (2012)

creation of a barrier against aluminium penetration into cells (Prabagar et al. 2011). Barceló et al. (1993) and Kidd et al. (2001) reported that silicon has increased concentration of root and shoot malate and phenolic compounds in aluminium treated maize (*Zea mays* L.). It has also been shown that silicon can ameliorate the plant morphology. Singh et al. (2011) reported that addition of silicon under aluminium exposure increased the frequency of stomatal and root length of rice (*Oryza sativa* L.) thus reducing aluminium toxicity symptoms. Silicon also ameliorates aluminium toxicity indirectly through improve phosphorus status of Al-treated wheat (*Triticum aestivum* L.) root epidermal and cortical cell walls (Ownby 1993).

As indicated above, silicon can ameliorate the negative effects of metal toxicity derived from aluminium toxicity by either internal (plant related) or external (soil related) mechanisms. Consequently, the productivity of key crops growing in acidic soils might thus be improved, and silicon could be used as a strategic tool to enhance plant tolerance to aluminium stress and to stimulate further research.

8.10 Aluminium and Boron Interaction

Boron is an essential element for plant growth. Soils of high aluminium concentrations are however characterized by high boron loses due to leaching (Hajiboland 2011). Root cells are the important site of the interactions between aluminium and boron since both aluminium toxicity and boron deficiency result in decreasing the rate of root growth (Yu and Goldbach 2007). Interactions between boron and aluminium is partly related to chemical similarities between them (Zhou et al. 2015).

Blevins and Lukaszewski (1998) claim that inhibition of plant aluminium toxicity as a result of boron depends on boron concentration in nutrient solution and plant species. For instance, boron supplementation reduced aluminium uptake by pea (*Pisum sativum* L.) roots and aluminium binding to cell walls, which resulted in less aluminium toxicity (Achary et al. 2008). Instead, Taylor and MacFie (1994) showed that boron did not alleviate aluminium toxicity symptoms in wheat (*Triticum aestivum* L.). One of the interaction of aluminium and boron is connected with cell wall. Boron is responsible for cross-link the pectic polysaccharide rhamnogalacturonan II (RG II), which results in a stable network of cell walls with decreased pore sizes and thus it may potentially reduces the access of aluminium to aluminium binding sites, thus tightening the wall (Heidarabadi et al. 2011). These results are coincident with the findings of Yang et al. (2004) on Soybean (*Glycine max* L.) seedlings. According to Yu and Goldbach (2007) these mechanisms stabilize apoplastic calcium decreasing its displacements from cell walls.

9 Conclusion

Efforts that aim to increase crop production depend, to a high degree, on the understanding of processes contributing to crop resistance to aluminium toxicity. Current levels of understanding of these processes have improved significantly in recent times, resulting in the breeding and engineering of new varieties of plants characterized by an increased resistance to aluminium. In contrast to laboratory experiments, where single stresses can usually be taken into account, crop reaction to aluminium toxicity is often altered by a range of soil properties and co-occurring stresses in field conditions. The resulting impact of these factors on plant function is not clearly understood and simple to predict as the combined effects of multiple stresses on various aspects of plant growth and function is often different from a simple sum of their actions. This may result in lower yields and increased costs of agricultural production. The research into aluminium toxicity to plants as affected by other environmental factors is imperative for a better understanding of the mechanisms of plant function in acid soils; this knowledge could be used to breed crops of increased resilience to predicted climate change and the associated increased probability of extreme weather events.

10 Summary

Crop plants are subjected to multiple abiotic stresses during their lifespan, which greatly reduce productivity and threaten global food security. How such interactions affect plant responses under multiple stresses, however, is less well understood, even though they may frequently occur in natural environments. In this review, we have summarized the interactions between plant responses to aluminium and co-stressors like drought, soil compaction, oxidative stress, heavy metals and nutrients imbalances at a cellular and morpho-physiological level.

As a result of the intensification of agriculture and the natural leaching processes, soil acidification is occurring more rapidly worldwide, this has the effect of increasing the aluminium concentration in soil, which raises the threat posed to both the environment and to human health. Aluminium toxicity induces diverse morphological, physiological and biochemical impairment of plant functioning, both directly or indirectly, and causes various harmful effects. In this review we underline the effect of aluminium on the plasma membrane, transport proteins, cell wall components and the DNA of root cells. We also highlight the defense mechanisms against aluminium including external processes (which limit the uptake of aluminium) and internal processes (detoxification mechanisms of aluminium that enters plant cells). The most frequently documented and the first consequence of aluminium toxicity in plant cells is root growth inhibition via the effect of aluminium at root apices cells. Nevertheless, the extent to which aluminium inhibits root growth depends on the plant species or the variety specific sensitivity and the concentration of these toxic ions in soil solutions.

Plant response to aluminium toxicity may be very difficult or even impossible to predict when it is accompanied by another stress factor, this is due to the complex interactions between the sites of activity (organs, cells), signalling, and the timing of single stresses. Aluminium reduces drought resistance by limiting the impact of water and nutrient uptake from deeper soil. Plant response is however, significantly affected by species or variety specific resistance to aluminium toxicity and the co-stressors which may exacerbate its effects. The interaction between aluminium and other abiotic stresses may be either additive or antagonistic. In general, the restricted rooting depth resulting from aluminium toxicity may be affected by the variability of the soil properties within the plant root system e.g. availability of water, nutrients and soil compaction which add another layer of complexity for consideration. Soil drought may reduce aluminium toxicity; this is an example of such an interaction. The inhibitory effect of aluminium on root growth is altered by the following effects of drought: decreasing root water uptake, increasing the concentration of aluminium in the soil solution, stimulating the effect of drought on downward root growth, increase of soil penetration resistance. Available case studies concerning drought-aluminium interactions indicate that drought may reduce aluminium toxicity due to reduced aluminium uptake. However, severe inhibition of root growth due to drought was observed even at low concentrations of aluminium in root tips. As to the above ground parts of the plant, drought was indicated as a factor reducing aluminium toxicity in spruce (*Picea abies* (L.) Karst).

Alteration in plant response to aluminium toxicity due to soil compaction, oxidative stress, lead and cadmium, the availability of soil minerals Mn, Fe, Ca, Mg, P, B, Si, and nitrogen forms NO_3^- and NH_4^+ were also reviewed.

Knowledge of the plant response to aluminium toxicity as affected by interactions with other abiotic stress factors is important for the selection of feasible methods aimed at the alleviation of the negative consequences for crop production in acidic soil and should be made more widely available.

References

Abd El-Azeem SAM, Ahmad M, Usman ARA, Kim KR, Oh SE, Lee SO, Ok YS (2013) Changes of biochemical properties and heavy metal bioavailability in soil treated with natural liming materials. Environ Earth Sci 70:3411–3420. doi:10.1007/s12665-013-2410-3

Achary VMM, Panda BB (2010) Aluminium-induced DNA damage and adaptive response to genotoxic stress in plant cells are mediated through reactive oxygen intermediates. Mutagenesis 25:201–209. doi:10.1093/mutage/gep063

Achary VMM, Jena S, Panda KK, Panda BB (2008) Aluminium induced oxidative stress and DNA damage in root cells of Allium cepa L. Ecotoxicol Environ Saf 70:300–310

Achary VMM, Parinandi NL, Panda BB (2013) Calcium channel blockers protect against aluminium-induced DNA damage and block adaptive response to genotoxic stress in plant cells. Mutat Res Genet Toxicol Environ Mutagen 751:130–138. doi:10.1016/j.mrgentox.2012.12.008

Ahn SJ, Sivaguru M, Osawa H, Chung GC, Matsumoto H (2001) Aluminium inhibits the H(+)-ATPase activity by permanently altering the plasma membrane surface potentials in squash roots. Plant Physiol 126:1381–1390

Alvarez I, Sam O, Reynaldo I, Testillano P, del Carmen RM, Arias M (2012) Morphological and cellular changes in rice roots (Oryza sativa L.) caused by Al stress. Bot Stud 53:67–73

Alvim MN, Ramos FT, Oliveira DC, Isaias R, França MGC (2012) Aluminium localization and toxicity symptoms related to root growth inhibition in rice (Oryza sativa L.) seedlings. J Biosci 37:1079–1088. doi:10.1007/s12038-012-9275-6

Barceló J, Guevara P, Poschenrieder C (1993) Silicon amelioration of aluminium toxicity in teosinte (Zea mays L. ssp. mexicana). Plant and Soil 154:249–255

Baylis AD, Gragopoulou C, Davidson KJ, Birchall JD (1994) Effects of silicon on the toxicity of aluminium to soybean. Commun Soil Sci Plant Anal 25:537–546

Bengough AG, McKenzie BM, Hallett PD, Valentine TA (2011) Root elongation, water stress, and mechanical impedance: a review of limiting stresses and beneficial root tip traits. J Exp Bot 62:59–68. doi:10.1093/jxb/erq350

Bhalerao SA, Prabhu DV (2013) Aluminium toxicity in plants—a review. J Appl Chem 2:447–474

Bhuja P, McLachlan K, Stephens J, Taylor G (2004) Accumulation of 1,3-beta-D-glucans, in response to aluminium and cytosolic calcium in Triticum aestivum *L.* Plant Cell Physiol 45:543–549. doi:10.1093/pcp/pch068

Bian M, Zhou M, Sun D, Li C (2013) Molecular approaches unravel the mechanism of acid soil tolerance in plants. Crop J 1(2):91–104. doi:10.1016/j.cj.2013.08.002

Blair LM, Taylor GJ (1997) The nature of interaction between aluminium and manganese on growth and metal accumulation in Triticum aestivum *L.* Environ Exp Bot 37:25–37. doi:10.1016/S0098-8472(96)01036-2

Blamey FPC (2001) The role of the root cell wall in aluminium toxicity. In: Ae N, Arihara J, Okada K, Srinivasan A (eds) Plant nutrient acquisition: new perspectives. Springer, Tokyo, pp 201–226

Blevins DG, Lukaszewski KM (1998) B in plant structure and function. Annu Rev Plant Physiol Plant Mol Biol 49:481–500

Bolan NS, Adriano DC, Curtin D (2003) Soil acidification and liming interactions with nutrient and heavy metal transformation. Adv Agron 78:215–269

Boscolo PRS, Menossi M, Jorge RA (2003) Aluminium-induced oxidative stress in maize. Phytochemistry 62:181–189. doi:10.1016/S0031-9422(02)00491-0

Bose J, Babourina O, Rengel Z (2011) Role of magnesium in alleviation of aluminium toxicity in plants. J Exp Bot 62:2251–2264. doi:10.1093/jxb/erq456

Brunner I, Sperisen C (2013) Aluminium exclusion and aluminium tolerance in woody plants. Front Plant Sci 4:1–12. doi:10.3389/fpls.2013.00172

Cai M, Wang N, Xing C, Wang F, Wu K, Du X (2013) Immobilization of aluminium with mucilage secreted by root cap and root border cells is related to aluminium resistance in Glycine max L. Environ Sci Pollut Res 20:8924–8933. doi:10.1007/s11356-013-1815-6

Cakmak I, Horst WJ (1991) Effect of aluminium on lipid peroxidation, superoxide dismutase, catalase and peroxidase activities in root tips of soybean (Glycine max L.). Physiol Plant 83:463–468

Cakmak I, Kirkby EA (2008) Role of magnesium in carbon partitioning and alleviating photooxidative damage. Physiol Plant 133:692–704. doi:10.1111/j.1399-3054.2007.01042.x

Cao Y, Lou Y, Han Y, Shi J, Wang Y, Ming F (2011) Al toxicity leads to enhanced cell division and changed photosynthesis in *Oryza rufipogon* L. Mol Biol Rep 38(8):4839–4846. doi:10.1007/s11033-010-0618-9

Chang YC, Ma JF, Matsumoto H (1998) Mechanisms of Al-induced iron chlorosis in wheat (*Triticum aestivum L.*). Al-inhibited biosynthesis and secretion of phytosiderophore. Plant Physiol 102(1):9–15. doi:10.1034/j.1399-3054.1998.1020102.x

Chen RF, Zhang FL, Zhang QM, Sun QB, Dong XY, Shen RF (2012) Aluminium-phosphorus interactions in plants growing on acid soils: does phosphorus always alleviate aluminium toxicity? J Sci Food Agric 92:995–1000. doi:10.1002/jsfa.4566
Choi CS, Sano H (2007) Abiotic-stress induces demethylation and transcriptional activation of a gene encoding a glycerophospho diesterase-like protein in tobacco plants. Mol Genet Genomics 277(5):589–600
Clarkson DT (1965) The effect of aluminium and some other trivalent metal cations on cell division in the root apices of *Allium cepa*. Ann Bot 29(2):310–315
Clarkson DT (1969) Metabolic aspects of aluminium toxicity and some possible mechanisms for resistance. In: Rorison IH (ed) Ecological aspects of the mineral nutrition of plants, British Ecological Symposium No. 9. Blackwell Scientific Publications, p 381–397
Clune TS, Copeland L (1999) Effects of aluminium on canola roots. Plant and Soil 216:27–33
Cumming JR, Ning J (2003) Arbuscular mycorrhizal fungi alter phosphorus relations of broomsedge (Andropogon virginicus L.) plants. J Exp Bot 54:1447–1459
da Silva IR, Ferrufino A, Sanzonowicz C, Smyth TJ, Israel DW, Carter TE Jr (2005) Interactions between magnesium, calcium, and aluminium on soybean root elongation. Rev Bras Ciênc Solo 29:747–754
de Camargo CDO, Filho AWPF (2001) Breeding bread wheat for tolerance to aluminium toxicity. In: Bedõ Z, Láng L (eds) Wheat in a global environment. Kluwer Academic, The Netherlands, pp 655–664
Delhaize E, Ryan PR (1995) Aluminium toxicity and tolerance in plants. Plant Physiol 107 (31):5–321
Delhaize E, Ryan PR, Hebb DM, Yamamoto Y, Sasaki T, Matsumoto H (2004) Engineering high-level aluminium tolerance in barley with the ALMT1 gene. Proc Natl Acad Sci U S A 101:15249–15254. doi:10.1073/pnas.0406258101
Domingues AM, Silva E, Freitas G, Ganança JF, Slaski JJ, Ângelo M, PDE C (2013) Aluminium tolerance in bean traditional cultivars from Madeira. Rev Ciênc Agron 36:148–156
Dong ZY, Wang YM, Zhang ZJ, Shen Y, Lin XY, Ou XF, Han FP, Liu B (2006) Extent and pattern of DNA methylation alteration in rice lines derived from introgressive hybridization of rice and Zizania latifolia Griseb. Theor Appl Genet 113:196–205
Dorneles AOS, Pereira AS, Rossato LV, Possebom G, Sasso VM, Bernardy K, Sandri RQ, Nicoloso FT, Ferreira PAA, Tabaldi LA (2016) Silicon reduces aluminium content in tissues and ameliorates its toxic effects on potato plant growth. Cienc Rural 46:506–512
Ezaki B, Katsuhara M, Kawamura M, Matsumoto H (2001) Different mechanisms of four aluminium (Al)-resistant transgenes for Al toxicity in Arabidopsis. Plant Physiol 127:918–927
Foy CD (1978) Plant adaptation to acid, aluminium-toxic soils. Commun Soil Sci Plant Anal 19:959–987
Gassmann W, Schroeder JI (1994) Inward-rectifying K1 channels in root hairs of wheat: a mechanism for aluminium-sensitive low-affinity K1 uptake and membrane potential control. Plant Physiol 105:1399–1408
Godbold DL, Kettner C (1991) Use of root elongation studies to determine aluminium and lead toxicity in Picea abies seedlings. J Plant Physiol 138(2):231–235
Goldman IL, Catrer TE Jr, Patterson RP (1989) A determinal interaction of subsoil aluminium and drought stress on the leaf water status of soybean. Agron J 81:461–463
Grauer UE, Horst WJ (1992) Modeling cation amelioration of aluminium phytotoxicity. Soil Sci Soc Am J 56:166–172. doi:10.2136/sssaj1992.03615995005600010026x
Gundersen P, Rasmussen L (1990) Nitrification in forest soils: effects from nitrogen deposition on soil acidification and aluminium release. Rev Environ Contam Toxicol 113:1–45
Guo TR, Zhang GP, Zhou MX, Wu FB, Chen JX (2004) Effects of aluminium and cadmium toxicity on growth and antioxidant enzyme activities of two barley genotypes with different Al resistance. Plant and Soil 258:241–248. doi:10.1023/B:PLSO.0000016554.87519.d6
Guo Z, Liao B, Huang C (2005) Mobility and speciation of Cd, Cu and Zn in two acidic soils affected by stimulated acid rain. J Environ Sci 17(2):332–334

Guo TR, Zhang GP, Zhou MX, Wu FB, Chen JX (2007) Influence of aluminium and cadmium stresses on mineral nutrition and root exudates in two barley cultivars. Pedosphere 17:505–512. doi:10.1016/S1002-0160(07)60060-5

Guo TR, Yao PC, Zhang ZD, Wang JJ, Wang M (2012) Involvement of antioxidative defense system in rice seedlings exposed to aluminium toxicity and phosphorus deficiency. Ric Sci 19:207–212. doi:10.1016/S1672-6308(12)60042-0

Gupta N, Gaurav SS, Kumar A (2013) Molecular basis of aluminium toxicity in plants: a review. Am J Plant Sci 4:21–37

Hajiboland R (2011) Effect of micronutrients deficiencies on plant stress responses. In: Ahmad P, Prasad MNV (eds) Abiotic stress responses in plants: metabolism, productivity and sustainability. Springer, New York

Haling RE, Simpson RJ, Culvenor RA, Lambers H, Richardson AE (2011) Effect of soil acidity, soil strength and macropores on root growth and morphology of perennial grass species differing in acid-soil resistance. Plant Cell Environ 34:444–456. doi:10.1111/j.1365-3040.2010.02254.x

Heidarabadi MD, Ghanati F, Fujiwara T (2011) Interaction between boron and aluminium and their effects on phenolic metabolism of Linum usitatissimum L. roots. Plant Physiol Biochem 49(12):1377–1383. doi:10.1016/j.plaphy.2011.09.008 Epub 21 Sept 2011

Hinsinger P (2001) Bioavailability of soil inorganic P in the rhizosphere as affected by root-induced chemical changes: a review. Plant and Soil 237:173–195

Hodson MJ, Evans DE (1995) Aluminium/silicon interactions in higher plants. J Exp Bot 46:161–171

Horst WJ (1995) The role of the apoplast in aluminium toxicity and resistance of higher plants: a review. Z Pflanzenernähr Bodenkd 158:419–428

Horst WJ, Wagner A, Marschner H (1982) Mucilage protects root meristems from aluminium injury. Z Pflanzenphysiol 109:95–103

Hossain AKMZ, Hossain MA, Koyama H, Hara T (2004) Effects of aluminium and boron supply on growth of seedlings among 15 cultivars of wheat (Triticum aestivum L.) grown in Bangladesh. Soil Sci Plant Nutr 50:189–195

Huang JW, Shaff JE, Grunes DL, Kochian LV (1992) Aluminium effects on calcium fluxes at the root apex of aluminium-tolerant and aluminium sensitive wheat cultivars. Plant Physiol 98:230–237

Huang J, Zhang Y, Peng JS, Zhong C, Yi HY, Ow DW, Gong JM (2012) Fission yeast HMT1 lowers seed cadmium through phytochelatin-dependent vacuolar sequestration in Arabidopsis. Plant Physiol 158:1779–1788. doi:10.1104/pp.111.192872

Huang W, Yang X, Yao S, Lwin OT, He H, Wang A, Li C, He L (2014) Reactive oxygen species burst induced by aluminium stress triggers mitochondria-dependent programmed cell death in peanut root tip cells. Plant Physiol Biochem 82:76–84. doi:10.1016/j.plaphy.2014.03.037

Ikeda H, Tadano T (1993) Ultrastructural changes of the root tip cells in barley induced by a comparatively low concentration of aluminium. Soil Sci Plant Nutr 39:109–117. doi:10.1080/00380768.1993.10416980

Jarvis SC, Hatch DJ (1986) The effects of low concentrations of aluminium on the growth and uptake of nitrate-N by white clover. Plant and Soil 95:43–55

Jerzykiewicz J (2001) Aluminium effect on nitrate assimilation in cucumber (Cucumis sativus L.) roots. Acta Physiol Plant 23:213–219. doi:10.1007/s11738-001-0011-3

Kidd PS, Llugany M, Poschenrieder C, Gunsé B, Barceló J (2001) The role of root exudates in aluminium resistance and silicon-induced ameliotarion of aluminium toxicity in three varieties of maize (Zea mays L.). J Exp Bot 52:1339–1352

Kimatu JN (2015) Correlating aluminium toxicity, heterosis and epigenetic mechanisms in maize yield improvement in acid soils. Biotechnol Mol Biol Rev 10:12–18. doi:10.5897/BMBR2014-0234

Kimatu JN, Diarso M, Song C, Agboola RS, Pang J (2011) DNA cytosine methylation alterations associated with aluminium toxicity and low pH in Sorghum bicolor. Afr J Agric Res 6:4579–4593. doi:10.5897/AJAR11.954

Kochian L (1995) Cellular mechanisms toxicity in plants. Annu Rev Plant Physiol Plant Mol Biol 46:237–260

Kopittke PM, Menzies NW, Blamey FPC (2005) Rhizotoxicity of aluminate and polycationic aluminium at high pH. Plant and Soil 266:177–186. doi:10.1007/s11104-005-2229-0

Kopittke PM, Blamey FPC, Menzies NW (2008) Toxicities of soluble Al, Cu, and La include ruptures to rhizodermal and root cortical cells of cowpea. Plant and Soil 303:217–227. doi:10.1007/s11104-007-9500-5

Lazarević B, Horvat T, Poljak M (2014) Effect of acid aluminous soil on photosynthetic parameters of potato (Solanum tuberosum L.). Potato Res 57:33–46. doi:10.1007/s11540-014-9251-7

Lee SE, Yim HK, Lim MN, Yoon IS, Kim JH, Hwang YS (2015) Abscisic acid prevents the coalescence of protein storage vacuoles by upregulating expression of a tonoplast intrinsic protein gene in barley aleurone. J Exp Bot 66(5):1191–1203. doi:10.1093/jxb/eru467

Liao H, Wan H, Shaff J, Wang X, Yan X, Kochian LV (2006) Phosphorus and aluminium interactions in soybean in relation to aluminium tolerance. Exudation of specific organic acids from different regions of the intact. Plant Physiol 141:674–684. doi:10.1104/pp.105.076497.674

Ligaba A, Maron L, Shaff J, Kochian L, Piñeros M (2012) Maize ZmALMT2 is a root anion transporter that mediates constitutive root malate efflux. Plant Cell Environ 35:1185–1200. doi:10.1111/j.1365-3040.2011.02479.x

Liu K, Luan S (2001) Internal aluminium block of plant inward K(+) channels. Plant Cell 13:1453–1465

Liu Q, Yang JL, He LS, Li YY, Zheng SJ (2008) Effect of aluminium on cell wall, plasma membrane, antioxidants and root elongation in triticale. Biol Plant 52:87–92. doi:10.1007/s10535-008-0014-7

Ma JF, Ryan PR (2010) Understanding how plants cope with acid soils. Funct Plant Biol 37:1–4

Manavalan LP, Guttikonda SK, Tran LSP, Nguyan HT (2009) Physiological and molecular approaches to improve drought resistance in soybean. Plant Cell Physiol 50:1260–1276. doi:10.1093/pcp/pcp082

Matsumoto H (2000) Cell biology of aluminium toxicity and tolerance in higher plants. Int Rev Cytol 200:1–47. doi:10.1016/S0074-7696(00)00001-2

Matsumoto H, Motoda H (2012) Aluminium toxicity recovery processes in root apices. Possible association with oxidative stress. Plant Sci 185–186:1–8. doi:10.1016/j.plantsci.2011.07.019

Matsumoto H, Yamaya T (1986) Inhibition of potassium and regulation of membrane-associated Mg^{2+}-ATPase activity of pea roots by aluminium. Soil Sci Plant Nutr 32:179–188. doi:10.1080/00380768.1986.10557495

Matsumoto H, Morimura S, Takahashi E (1977) Binding of aluminium to DNA of DNP in pea root nuclei. Plant Cell Physiol 18:987–993

Meriño-Gergichevich C, Alberdi M, Ivanov AG, Reyes-Diaz M (2010) Al^{3+}–Ca^{2+} interaction in plants growing in acid soils: Al-phytotoxicity response to calcareous amendments. J Soil Sci Plant Nutr 10:217–243

Milla MAR, Butler E, Huete AR, Wilson CF, Anderson O, Gustafson JP (2002) Expressed sequence tag-based gene expression analysis under aluminium stress in rye. Plant Physiol 130:1706–1716

Nezames CD, Sjogren CA, Barajas JF, Larsen PB (2012) The Arabidopsis cell cycle checkpoint regulators TANMEI/ALT2 and ATR mediate the active process of aluminium-dependent root growth inhibition. Plant Cell 24:608–621. doi:10.1105/tpc.112.095596 Epub 17 Feb 2012

Osawa H, Endo I, Hara Y, Matsushima Y, Tange T (2011) Transient proliferation of proanthocyanidin-accumulating cells on the epidermal apex contributes to highly aluminium-resistant root elongation in camphor tree. Plant Physiol 155(1):433–446. doi:10.1104/pp.110.166967

Ownby JD (1993) Mechanisms of reaction of hematoxylin with aluminium-treated wheat roots. Physiol Plant 87:371–380
Palmgren MG, Harper JF (1999) Pumping with plant P-type ATPases. J Exp Bot 50:883–893
Pal'ove-Balang P, Mistrik I (2007) Impact of low pH and aluminium on nitrogen uptake and metabolism in roots of Lotus japonicus. Biologia 62:715–719. doi:10.2478/s11756-007-0133-1
Pal'ove-Balang P, Čiamporová M, Zelinová V, Pavlovkin J, Gurinová E, Mistrík I (2012) Cellular responses of two Latin-American cultivars of Lotus corniculatus to low pH and Al stress. Cent Eur J Biol 7:1046–1054. doi:10.2478/s11535-012-0098-0
Panda SK, Matsumoto H (2007) Molecular physiology of aluminium toxicity and tolerance in plants. Bot Rev 73:326–347. doi:10.1663/0006-8101(2007)73[326:MPOATA]2.0.CO;2
Panda SK, Yamamoto Y, Kondo H, Matsumoto H (2008) Mitochondrial alterations related to programmed cell death in tobacco cells under aluminium stress. C R Biol 331:597–610. doi:10.1016/j.crvi.2008.04.008
Pécsváradi AZ, Nagy Z, Varga A, Vashegyi A, Labádi I, Galbács G, Zsoldos F (2009) Chloroplastic glutamine synthetase is activated by direct binding of aluminium. Physiol Plant 135:43–50
Pina RG, Cervantes C (1996) Microbial interactions with aluminium. Biometals 9:311–316
Pontigo S, Ribera A, Gianfreda L, de la Luz MM, Nikolic M, Cartes P (2015) Silicon in vascular plants: uptake, transport and its influence on mineral stress under acidic conditions. Planta 242:23–37. doi:10.1007/s00425-015-2333-1
Poschenrieder C, Gunsé B, Corrales I, Barceló J (2008) A glance into aluminium toxicity and resistance in plants. Sci Total Environ 400:356–368. doi:10.1016/j.scitotenv.2008.06.003
Prabagar S, Hodson MJ, Evans DE (2011) Silicon amelioration of aluminium toxicity and cell death in suspension cultures of Norway spruce (Picea abies (L.) Karst). Environ Exp Bot 70:266–276. doi:10.1016/j.envexpbot.2010.10.001
Purcino AAC, Carvalho Alves VM, Parentoni SN, Belele ChL, Loguercio LL (2003) Aluminium effects on nitrogen uptake and nitrogen assimilating enzymes in maize genotypes with contrasting tolerance to aluminium toxicity. J Plant Nutr 21(1):31–61. doi:10.1081/PLN-120016496
Qifu MA, Rengel Z, Kuo J (2002) Aluminium toxicity in rye (*Secale cereale*): root growth and dynamics of cytoplasmic Ca^{2+} in intact root tips. Ann Bot 89:241–244
Rengel Z (1996) Tansley review no 89—uptake of aluminium by plant cells. New Phytol 134:389–406
Rengel Z, Elliott DC (1992) Aluminium inhibits net $45Ca^{2+}$ uptake by Amaranthus protoplasts. Biochem Physiol Pflanz 188:177–186
Rengel Z, Robinson DL (1989) Competitive Al^{3+} inhibition of net Mg^{2+} uptake by intact lolium multiflorum roots: I. Kinetics. Plant Physiol 91:1407–1413. doi:10.1104/pp.91.4.1407
Rengel Z, Zhang WH (2003) Role of dynamics of intracellular calcium in aluminium-toxicity syndrome. New Phytol 159:295–314. doi:10.1046/j.1469-8137.2003.00821.x
Rounds MA, Larsen PB (2008) Aluminium-dependent root-growth inhibition in Arabidopsis results from AtATR-regulated cell-cycle arrest. Curr Biol 18:1495–1500. doi:10.1016/j.cub.2008.08.050
Rouphael Y, Cardarelli M, Colla G (2015) Role of arbuscular mycorrhizal fungi in alleviating the adverse effects of acidity and aluminium toxicity in zucchini squash. Sci Hortic 188:97–105. doi:10.1016/j.scienta.2015.03.031
Rufty TW Jr, MacKown CT, Lazof DB, Carter TE (1995) Effects of aluminium on nitrate uptake and assimilation. Plant Cell Environ 18(11):1325–1331
Ryan PR, Ditomaso JM, Kochian LV (1993) Aluminium toxicity in roots: an investigation of spatial sensitivity and the role of the root cap. J Exp Bot 44:437–446. doi:10.1093/jxb/44.2.437
Samac DA, Tesfaye M (2003) Plant improvement for tolerance to aluminium in acid soils—a review. Plant Cell Tiss Org Cult 75:189–207. doi:10.1023/A:1025843829545
Sampson M, Clarkson D, Davies DD (1965) DNA synthesis in aluminium-treated roots of barley. Science 148:1476–1477

Sasaki T, Yamamoto Y, Ezaki B, Katsuhara M, Ahn SJ, Ryan PR, Delhaize E, Matsumoto H (2004) A wheat gene encoding an aluminium-activated malate transporter. Plant J 37:645–653. doi:10.1111/j.1365-313X.2003.01991.x

Scheffer-Basso SM, Prior BC (2015) Aluminium toxicity in roots of legume seedlings assessed by topological analysis. Acta Sci Agron 37:61–68. doi:10.4025/actasciagron.v37i1.18362

Schier GA, McQuattie CJ (2000) Effect of water stress on aluminium toxicity in pitch pine seedlings. J Plant Nutr 23:637–647

Schmohl N, Horst WJ (2000) Cell wall pectin content modulates aluminium sensitivity of Zea mays (L.) cells grown in suspension culture. Plant Cell Environ 23:735–742. doi:10.1046/j.1365-3040.2000.00591.x

Scholl LV, Keltjens WG, Hoffland E, Breemen NV (2005) Effect of ectomycorrhizal colonization on the uptake of Ca, Mg, and Al by *Pinus sylvestris* under aluminium toxicity. For Ecol Manage 215:352–360. doi:10.1016/j.foreco.2005.05.025

Shamsi IH, Wei K, Jilani G, Zhang G (2007) Interactions of cadmium and aluminium toxicity in their effect on growth and physiological parameters in soybean. J Zhejiang Univ Sci B 8:181–188. doi:10.1631/jzus.2007.B0181

Silva IR, Smyth TJ, Moxley DF, Carter TE, Allen NS, Rufty TW (2000) Aluminium accumulation at nuclei of cells in the root tip. Fluorescence detection using lumogallion and confocal laser scanning microscopy. Plant Physiol 123:543–552. doi:10.1104/pp.123.2.543

Silva IR, Ferrufino A, Sanzonowicz C, Smyth TJ, Israel DW, Carter TE Jr (2005) Interactions between magnesium, calcium, and aluminium on soybean root elongation. Rev Bras Ciênc Solo 29:747–754. doi:10.1590/S0100-06832005000500010

Silva S, Pinto-Carnide O, Martins-Lopes P, Matos M, Guedes-Pint H, Santos C (2010) Differential aluminium changes on nutrient accumulation and root differentiation in an Al sensitive vs. tolerant wheat. Environ Exp Bot 68:91–98. doi:10.1016/j.envexpbot.2009.10.005

Simões CC, Melo JO, Magalhaes JV, Guimarães CT (2012) Genetic and molecular mechanisms of aluminium tolerance in plants. Genet Mol Res 11:1949–1957. doi:10.4238/2012.July.19.14

Singh VP, Tripathi DK, Kumar D, Chauhan DK (2011) Influence of exogenous silicon addition on aluminium tolerance in rice seedlings. Biol Trace Elem Res 144:1260–1274

Singh S, Verma A, Dubey VK (2012) Effectivity of anti-oxidative enzymatic system on diminishing the oxidative stress induced by aluminium in chickpea (Cicer arietinum L.) seedlings. Braz J Plant Physiol 24:47–54. doi:10.1590/S1677-04202012000100007

Sivaguru M, Fujiwara T, Samaj J, Baluska F, Yang ZM, Osawa H, Maeda T, Mori T, Volkmann D, Matsumoto H (2000) Aluminium-induced 1→3-β-D-glucan inhibits cell-to-cell trafficking of molecules through plasmodesmata. A new mechanism of aluminium toxicity in plants. Plant Physiol 124:991–1005

Slugeňová K, Ditmarová Ľ, Kurjak D, Vaľka J (2011) Drought and aluminium as stress factors in Norway spruce (Picea abies [L.] Karst) seedlings. J For Sci 57:547–554

Sun Q, Bin SRF, Zhao XQ, Chen RF, Dong XY (2008) Phosphorus enhances Al resistance in Al-resistant Lespedeza bicolor but not in Al-sensitive L. cuneata under relatively high Al stress. Ann Bot 102:795–804. doi:10.1093/aob/mcn166

Tamás L, Huttová J, Mistrík I, Simonovicová M, Siroká B (2006) Aluminium-induced drought and oxidative stress in barley roots. J Plant Physiol 163:781–784. doi:10.1016/j.jplph.2005.08.012

Tan K, Keltjens WG (1990) Interaction between aluminium and phosphorus in sorghum plants. Plant and Soil 23:15–23. doi:10.1007/BF00010927

Tan K, Keltjens WG (1995) Analysis of acid-soil stress in sorghum genotypes with emphasis on aluminium and magnesium interactions. Plant and Soil 171:147–150

Taylor GJ (1988) Mechanisms of aluminium tolerance in *Triticum aestivum* (wheat). V. Nitrogen nutrition, plant-induced pH and tolerance to aluminium: correlation without causality? Can J Bot 66:694–699

Taylor WK, MacFie SM (1994) Modelling the potential for B amelioration of aluminium toxicity using the Weibull function. Can J Bot 72:1187–1196

Tesfaye M, Temple SJ, Allan DL, Vance CP, Samac DA (2001) Overexpression of malate dehydrogenase in transgenic alfalfa enhances organic acid synthesis and confers tolerance to aluminium. Plant Physiol 127:1836–1844. doi:10.1104/pp.010376

Too EJ, Carlsson AS, Onkware AO, Were BA, Geleta M, Bryngelsson T, Gudu S (2014) Cell membrane integrity, callose accumulation, and root growth in aluminium-stressed sorghum seedlings. Biol Plant 58:768–772. doi:10.1007/s10535-014-0455-0

Vardar F, Ar E, Gözük N (2006) Effects of aluminium on in vitro root growth and seed germination of tobacco (Nicotiana tabacum L.). Adv Food Sci 28:85–88

Wahyudi I, Handayanto E (2015) The potential of legume tree prunings as organic matters for improving phosphorus availability in an acid soil. J Degrad Min Lands Manage 2:259–266. doi:10.15243/jdmlm.2014.022.259

Wang Y, Stass A, Horst WJ (2004) Apoplastic binding of aluminium is involved in silicon-induced amelioration of aluminium toxicity in maize. Plant Physiol 136:3762–3770. doi:10.1104/pp.104.045005

Wang J, Raman H, Zhang G, Mendham N, Zhou M (2006) Aluminium tolerance in barley (Hordeum vulgare L.): physiological mechanisms, genetics and screening methods. J Zhejiang Univ Sci B 7:769–787. doi:10.1631/jzus.2006.B0769

Wang TZ, Tian QY, Wang BL, Zhao MG, Zhang WH (2014) Genome variations account for different response to three mineral elements between Medicago truncatula ecotypes Jemalong A17 and R108. BMC Plant Biol 14:1–11. doi:10.1186/1471-2229-14-122

Wang W, Zhao XQ, Chen RF, Dong XY, Lan P, Ma JF, Shen RF (2015) Altered cell wall properties are responsible for ammonium-reduced aluminium accumulation in rice roots. Plant Cell Environ 38:1382–1390. doi:10.1111/pce.12490

Watanabe T, Jansen J, Osaki M (2006) Al–Fe interactions and growth enhancement in *Melastoma Malabathricum* and *Miscanthus sinensis* dominating acid sulphate soils. Plant Cell Environ 29:2124–2132

Watanabe T, Misawa S, Hiradate S, Osaki M (2008) Root mucilage enhances aluminium accumulation in Melastoma malabathricum, an aluminium accumulator. Plant Signal Behav 3:603–605. doi:10.1111/j.1469-8137.2008.02397.x

White PJ, Broadley MR (2003) Calcium in plants. Ann Bot 92:487–511

Whitmore AP, Whalley WR, Bird NR, Watts CW, Gregory AS (2011) Estimating soil strength in the rooting zone of wheat. Plant and Soil 339:363–375. doi:10.1007/s11104-010-0588-7

Wu FB, Zhang GP (2002) Genotypic variation in kernel heavy metal concentrations in barley and as affected by soil factors. J Plant Nutr 25(6):1163–1173. doi:10.1081/PLN-120004380

Xia J, Yamaji N, Kasai T, Ma JF (2010) Plasma membrane-localized transporter for aluminium in rice. Proc Natl Acad Sci U S A 107:18381–18385. doi:10.1073/pnas.1004949107

Yamamoto Y, Kobayashi Y, Matsumoto H (2001) Lipid peroxidation is an early symptom triggered by aluminium, but not the primary cause of elongation inhibition in pea roots. Plant Physiol 125:199–208. doi:10.1104/pp.125.1.199

Yang YH, Gu HJ, Fan WY, Bilkisu A (2004) Effects of boron on aluminium toxicity on seedlings of two soybean cultivars. Water Air Soil Pollut 154:239–248

Yang JL, Li YY, Zhang YJ, Zhang SS, Wu YR, Wu P, Zheng SJ (2008) Cell Wall polysaccharides are specifically involved in the exclusion of aluminium from the rice root apex. Plant Physiol 146:602–611. doi:10.1104/pp.107.111989

Yang ZB, Eticha D, Rotter B, Rao IM, Horst WJ (2011) Physiological and molecular analysis of polythylene glycol-induced reduction of aluminium accumulation in the root tips of common bean (Phaseolus vulgaris). New Phytol 192:99–113. doi:10.1111/j.1469-8137.2011.03784.x

Yu M, Goldbach HE (2007) Influence of boron on Al absorption and Ca release of root border cells of pea (Pisum sativum). In: Xu F, Goldbach HE, Brown PH, Bell RW, Fujiwara T, Hunt CD, Goldberg S, Shi L (eds), Advances in plant and animal boron nutrition. Proceedings of the 3rd international symposium on all aspects of plant and animal boron nutrition, pp 63–68. ISBN: 978-1-4020-5381-8

Zhang WH, Rengel Z (1999) Aluminium induces an increase in cytoplasmic calcium in intact wheat root apical cells. Aust J Plant Physiol 26:401–409

Zhang H, Jiang Z, Qin R, Zhang H, Zou J, Jiang W, Liu D (2014) Accumulation and cellular toxicity of aluminium in seedling of Pinus massoniana. BMC Plant Biol 14:1–16. doi:10.1186/s12870-014-0264-9

Zhang YK, Zhu DF, Zhang YP, Chen HZ, Xiang J, Lin XQ (2015) Low pH-induced changes of antioxidant enzyme and ATPase activities in the roots of rice (Oryza sativa L.) seedlings. PLoS One 10:1–12. doi:10.1371/journal.pone.0116971

Zhao Z, Ma JF, Sato K, Takeda K (2003) Differential Al resistance and citrate secretion in barley (Hordeum vulgare L.). Planta 217:794–800. doi:10.1007/s00425-003-1043-2

Zheng SJ (2010) Crop production on acidic soils: overcoming aluminium toxicity and phosphorus deficiency. Ann Bot 106:183–184. doi:10.1093/aob/mcq134

Zheng SJ, Yang JL, He YF, Yu XH, Zhang L, You JF, Shen RF, Matsumoto H (2005) Immobilization of aluminium with phosphorus in roots is associated with high aluminium resistance in buckwheat. Plant Physiol 138:297–303. doi:10.1104/pp.105.059667

Zheng L, Lan P, Shen RF, Li WF (2014) Proteomics of aluminium tolerance in plants. Proteomics 14:566–578. doi:10.1002/pmic.201300252

Zhou XX, Yang LT, Qi YP, Guo P, Chen LS (2015) Mechanisms on boron-induced alleviation of aluminium-toxicity in Citrus grandis seedlings at a transcriptional level revealed by cDNA-AFLP analysis. PLoS One 10(3):e0115485. doi:10.1371/journal.pone.0115485

Zhu XF, Shi YZ, Lei GJ, Fry SC, Zhang BC, Zhou YH et al (2012) XTH31, encoding an in vitro XEH/XET-active enzyme, regulates aluminium sensitivity by modulating in vivo XET action, cell wall xyloglucan content, and aluminium binding capacity in Arabidopsis. Plant Cell 24:4731–4747. doi:10.1105/tpc.112.106039

ARSENIC: A Review on Exposure Pathways, Accumulation, Mobility and Transmission into the Human Food Chain

Beste Arslan, Mustafa B.A. Djamgoz, and Ertan Akün

Contents

1 Introduction 27
2 Pathways of Exposure 28
3 Transmission and Mobility of Arsenic in Environment 31
4 Transfer from Soil-Water to Plants 34
5 Arsenic in Plants 36
6 Arsenic in Rice and Effect of Processing 39
7 Conclusion 45
8 Summary 46
References 47

1 Introduction

Arsenic (As) contamination of the environment from both natural and anthropogenic sources is nowadays a major environmental concern in various parts of the world, due to its persistence and carcinogenic effects on living organisms at certain concentrations. It is estimated that more than 150 million people around the world are exposed to As, consuming water containing high levels of As (Ravenscroft et al. 2009). Besides drinking water, diet also plays critical role in As

B. Arslan (✉) • E. Akün
Faculty of Engineering, Biotechnology Research Centre, Cyprus International University, Haspolat, Mersin 10, Nicosia, Northern Cyprus, Turkey
e-mail: barslan@ciu.edu.tr; eakun@ciu.edu.tr

M.B.A. Djamgoz
Faculty of Engineering, Biotechnology Research Centre, Cyprus International University, Haspolat, Mersin 10, Nicosia, Northern Cyprus, Turkey

Division of Cell and Molecular Biology, Neuroscience Solutions to Cancer Research Group, Imperial College London, Sir Alexander Fleming Building, South Kensington Campus, London SW7 2AZ, UK
e-mail: m.djamgoz@imperial.ac.uk

P. de Voogt (ed.), *Reviews of Environmental Contamination and Toxicology Volume 243*, Reviews of Environmental Contamination and Toxicology 243,
DOI 10.1007/398_2016_18

exposure. Vegetables and crops grown in As contaminated soil and irrigation water can take up As through their roots and accumulate it in their edible tissues, which are then consumed by people or animals.

It is a well-known fact that arsenic toxicity is evaluated by considering its presence, nature (organic-inorganic) and concentration in different media such as soil, water and plants. Presence of As in soils varies with As mobility, soil properties and adsorption/desorption capacity of minerals that occur together with As (Kar et al. 2010; Hsu et al. 2012). Each person is affected by As at different levels at similar concentrations depending on age, genetic characteristics, type of exposure (acute or chronic) and presence of co-morbidities (Keil et al. 2011). In addition to these, duration and magnitude of exposure, origin of As and daily dietary habits are also important factors, determining the severity of the effects (Kapaj et al. 2006).

Due to the fact that arsenic does not have taste, odor and color even at high concentrations, it is neither easy to detect its presence in water and food, nor is it easy to avoid exposure to it (Smith and Steinmaus 2009). This is the main reason why additional detailed studies are necessary to avoid exposure to it, such as its transfer capabilities in different media. Metals such as Cd, Zn, Cu including As are persistent in the environment, unlike organic chemicals which act on the contrary (Keil et al. 2011). However, once As is present in the environment, it can easily spread from soil—water, water—soil and then to plants, due to its high transfer capacity. In addition to these, it can transfer from air to soil via dust and atmospheric deposition, whereas it can as well transfer to surface and groundwater through precipitation. Through dust transfer, it can be transported over long distances attached to suspended particles in air. Csavina et al. (2012) discussed the strong ability of ultra-fine particles formed from mine tailings to disperse and flocculate in the range of 0.1–1 μm particle size in the atmosphere. Even though many people are of the belief that soil and water are the major transporting agents for heavy metals, it is observed that air is also a significant transporting agent for As and future researches should concentrate more on this issue.

According to Chung et al. (2014), As can be dissolved in water causing contamination of both surface and ground waters. It has the ability of changing its valence, attaching to or separating from particles that exist in sediments, water or air, in case of interaction with oxygen or other molecules.

2 Pathways of Exposure

Arsenic (As) occurs in all environmental compartments (water, soil and air) and can be taken up by animals and plants and consequently transferred to humans. *Ingestion* of As containing drinking water and foods; *dermal uptake* of As contaminated soil/sediment; *inhalation* of suspended arsenical dusts in the air are some of the pathways of As exposure (Lijzen et al. 2001; Orloff et al. 2009; Biswas et al. 2012). In some cases, drinking of groundwater containing high levels of naturally occurring inorganic As may lead to mass poisoning, since it is a direct exposure pathway. This is how and why Smith et al. (2000) reported an incident at Bangladesh in 1996 as the largest mass poisoning, where more than half of the population of the country were endangered.

Total dietary exposure to As should be calculated considering both drinking water and foodstuffs. However, most of the research reported in literature covers only As amount in drinking water as the total exposure parameter, causing underestimation of health risks. For people who are not exposed to As through drinking water; food might be the main contributor of As exposure. Figure 1 presents all the mechanisms/pathways involved in As exposure of humans.

Soils, sediments, groundwater, surface water and air are the environments for accumulating As. This As accumulated in these environments is then transmitted to plants, animals and humans through the food chain. Due to the fact that toxic heavy metals including As persist in the environment for a long time, it becomes a serious concern due to their probable harmful effects. Both direct and indirect pathways have great effect on human exposure (Lam and Sia Su 2009). One of the most dangerous situations occurs when As contaminated groundwater is continuously used for irrigating vegetables and crops. Under such circumstances, irrigation with As enriched groundwater becomes the main pathway for As to enter the human body (Samal et al. 2011; Chatterjee et al. 2010).

Presence of As in surface soils is either natural or artificial. Natural presence stands for As already present within the soils, while artificial presence stands for As introduced by pesticides, herbicides and As contaminated irrigation waters for long periods (Huang et al. 2006; Bhattacharya et al. 2010). Ramirez-Andreotta et al. (2013) demonstrated a linear correlation between As concentration in soils and accumulated As amount in edible parts of some vegetables (lettuce, radish,

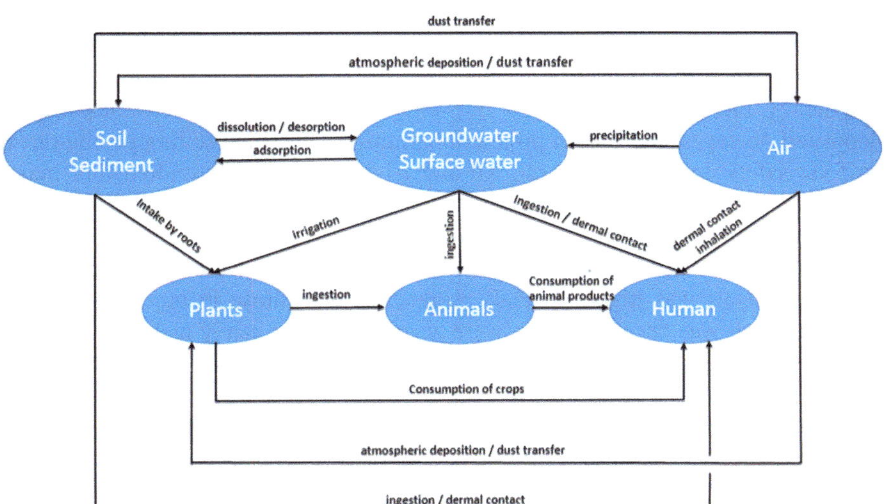

Fig. 1 Pathways of As exposure

broccoli, cabbage, onion, garlic, spinach, beets and bean). The study did not show any correlation for the following vegetables: tomato, pepper, cucumber and squash. Probably because, these vegetables have less uptake capacity compared to others so that they do not accumulate high amounts of As even at high levels of bioavailability in soil. The study thus differentiated bio-concentration factors (BCF) for different vegetable species, where non-correlation findings depend on As absorption capacity of root structure, as well as genetic and physiological adaptations which allow them to absorb, translocate from root to shoot and resist high amount of As in their cells (Wang et al. 2009; Bondada et al. 2007; Samal et al. 2011).

Sometimes, anthropogenic activities affect the toxic heavy metal content of agricultural soils, such as mining and metallurgical activities. The As present in the contaminated soils then may enter the edible parts of vegetables creating health risks for the inhabitants surrounding the area. The spread of contamination is not limited to the nearby surroundings and under certain circumstances such as winds and surface water runoff during excess rainfalls, contamination spreads to long distances. Samples of alfalfa, onion and carrot collected at Vinto-Oruro mining district illustrated high contents of As in these vegetables at concentrations of 399, 99 and 92 ppm respectively. Same enlargement mechanism for airborne transportation is valid for metallurgical regions as well. A decreasing trend of As accumulation in edible parts with respect to the distance from the metallurgical zone was also noted, which suggested that airborne transport of As was the basic reason for high As accumulation in different vegetables (Mercado et al. 2009). In another study carried out around the Zambian Copper belt, the cassava leaves studied did not indicate any excessive Cu ingestion but Pb and As contamination was occasionally recorded. The dust coming from smelters was the major source of contamination (Kříbek et al. 2014). Similarly, numerous studies on human urine and blood demonstrated the inverse relationship of As exposure with distance of inhabited areas from metal smelters or mining areas. This situation once again illustrates the long distance travel of airborne As with convenient wind speed and direction (Csavina et al. 2012). On the other hand, we must bear in mind the fact that heavy metals in air, emitted by vehicles and industrial activities can be deposited on vegetables during production, transportation and selling posing threat on food safety and human health. Ali and Al-Qahtani (2012) studied four major industrial areas in Saudi Arabia (Tabouk, Riyadh, Damamm and Jazan) and searched for the accumulation of heavy metals in different parts of the vegetables (roots, stems, leaves, fruits, cereals and legumes). They found that the accumulation of heavy metals was higher in leafy vegetables than those in other parts. They claimed that it was because leaves were access points of heavy metals from air to vegetables. Also, epidemiological studies carried out indicated that ingested and inhaled As can cause skin and lung cancers respectively. Some other studies also showed that As ingestion may as well lead to internal cancers (Phan et al. 2013).

In Beijing-China, a large scale study was carried out in 2006 to analyze risks of high As concentrations in vegetables to human health. Samples were collected from fields, greenhouses and as well as from supermarkets. A HG-AFS method was adopted to determine concentrations and the results indicated presence of As in soils in the range 4.44–25.3 ppm, which can be regarded as a significant contamination. In different regions, the maximum permissible concentration of As in vegetables was exceeded (Shandong District: 0.331 ppm vs. 0.25 ppm), while in some this concentration was less (Fengtai District: 0.479 ppm vs. 0.5 ppm). It was also observed that As concentration was much more in field-grown vegetables when compared with those planted in greenhouses (Chen et al. 2006). This might be because vegetables in greenhouses are not affected by air deposition and anthropogenic activities in the surroundings, as much as the vegetables that are grown in the open fields.

The study also indicated a significant difference with respect to bio-concentration factor (BCF):

a. Group of vegetables with higher BCF's: Rape, radish, pakchoi, onion, mustard, cucumber, Chinese cabbage and cabbage
b. Group of vegetables with lower BCF's: chili, beans, wax gourd, eggplant, spinach, tomato and celery

3 Transmission and Mobility of Arsenic in Environment

The presence of arsenic (As) in aquifer sediments can be the result of various biogeochemical processes, such as *reductive dissolution of As bearing minerals*, *oxidation hypothesis*, *desorption in alkaline environments*, and *geothermal influence* (Yadav et al. 2015; Chauhan et al. 2009). Detailed explanations regarding these processes are presented below:

a. *Reductive dissolution of As bearing minerals*: For a long time, reductive dissolution of As bearing minerals was believed to be the main mechanism of As release through weathering of As bearing minerals such as arsenopyrite and other sulfide minerals. Goldberg and Johnston (2001) discussed the well-known association between As and Fe(III) oxides in soils and sediments, since Fe oxides and oxyhydroxides are the most significant absorbents for As due to their high binding capacity. Reductive dissolution of As-bearing, amorphous iron oxide/hydroxide minerals is the most common cause of As mobilization into aquifer systems in reducing environments, where As(V) is promoted to reduce As(III) and to release into groundwater in anaerobic conditions (Islam et al. 2004; Ramos et al. 2014).
b. *Oxidation hypothesis*: When water table drawdowns and oxygen invades the aquifer, As rich sulfide minerals get oxidized and As is released into the groundwater (Das et al. 1996). Although this is a well-known hypothesis, Yadav et al. (2014) claimed that the oxidation of As rich minerals in Nawalparasi District-Nepal did not cause any change in groundwater As concentration between dry and wet season. Rather, As was released due to the reductive dissolution mechanism of Fe (III) oxyhydroxides in a reducing environment. Studies in a mining district at Zimapan, in the Hidalgo province of

central Mexico, showed that, oxidation and dissolution of As containing minerals within a fractured limestone aquifer is the main As release mechanism, in addition to leaching of As from mine tailings (Sracek et al. 2010; Ongley et al. 2007). Besides these mechanisms, it is suggested that As is desorbed from the surface of hydrous metal oxides and clay minerals under alkaline ($pH > 8$) and oxidative conditions and released into ambient groundwater (Smedley and Kinniburgh 2002; Bhattacharya et al. 2006).

c. *Desorption in alkaline environments*: Rango et al. (2013) demonstrated the importance of elevated groundwater pH on mobility of As and other natural occurring contaminants from Quaternary sedimentary aquifers of the Main Ethiopian Rift (MER). This suggests that the activity of As depends on alkaline desorption processes. According to results obtained the highest concentration of As was determined in waters with pH values ranging between 8.1 and 8.9. Similarly, Masscheleyn et al. 1991 showed that desorption of arsenate was enhanced with higher pH values.
d. *Geothermal influence*: Geothermal water rising through a fractured zone can mix with "cold" shallow aquifers and surface waters causing contamination (Aksoy et al. 2009). In addition, besides mixing of geothermal waters, waste water derived from geothermal plants can also lead to contamination of shallow groundwater and surface water by As and other toxic elements (Gemici and Tarcan 2004; Demirel and Yıldırım 2002). This is confirmed by Bundschuh et al. (2013) in a study area in Turkey. If reductive dissolution of As occurs near a geothermal area, mobility of As is induced by high temperature and pressure of liquid and high residence time. Observed in different countries including Turkey, India, Bangladesh and Taiwan, the tectonic activities at or near geothermal deposits are responsible for the increased temperature and pressure. On the other hand, mobilization of arsenic present in geothermal fluids depends on the time of residence. When the residence time increases, mobilization of As will continue under the present supportive conditions. Geothermal waters may contain both As (III) and As(V) species, however as the water rises up and has contact with atmospheric oxygen near earth surface, As(III) is oxidized to As(V) and dissolved redox sensitive minerals such as Fe oxides are precipitated and change their oxidation states (Webster and Nordstrom 2003; Alsina et al. 2007). Bundschuh et al. (2013) carried out a study in geothermal waters in deep wells and hot springs in western Anatolia and observed that, negative Eh values and positive Eh values correspond to dominance of As(III) and As(V), respectively. This also indicates that As III which is widespread in geothermal waters is formed in reducing environments, while As V is formed in oxidation environments.

It is important to understand the hydrogeological and hydrochemical impacts on As distribution to be able to assess As mobility in groundwater systems. Transmission and mobility of As in the environment affects As concentration. This mobilization is also effective in the case of interrelationship of As with hydro-geochemical characteristics of groundwater. Geological and geomorphological structures are not constant on the earth's surface and may laterally vary to a great extent, even in a small agricultural zone. For instance, as the hydraulic conductivity of aquifer sediments gets lower and surface topography is flatter, groundwater is mostly immobile (Guo et al. 2008). In the same way, composition and matrix of groundwater, interaction between surface and groundwater, reduction potential, fluctuations in groundwater level, recharge rate, distribution of dissolved oxygen, and organic matter may also vary (Yadav et al. 2015). Depth of the water table affects As concentrations, because of two main reasons. First, an increase in water table brings the groundwater table closer to the land surface and mixes groundwater with agrochemical and other wastes at or near the surface. Second, the rise in water table will cause dissolution of Fe-oxyhydroxide; giving way to the release of As under reducing environmental conditions (Bhattacharya et al. 2001; Nickson et al. 1998). Microbes, on the other hand, can catalytically decompose organic compound resulting in release of Fe, Mn and HCO_3 by the dissolving of As-rich Fe and Mn-oxyhydroxide (Kar et al. 2010; Liaoa et al. 2011; Nickson et al. 1998; Yan et al. 2000). The leachable As content was also observed to be high in organic matter phase, which indicated the important role of microbial populations and organic matter in the mobility of As under reducing condition. As for the reduction potential, high redox conditions with the induction of high concentrations of Fe and Mn enhance elevated levels of As. During reductive dissolution mechanism, As gets adsorbed onto Fe (III)-oxyhydroxides and is released under reduction condition (Islam et al. 2004; Nickson et al. 1998). Moreover, dissolved oxygen in groundwater is consumed during microbial oxidation, inducing an increase in HCO_3 concentration (Liaoa et al. 2011; Kar et al. 2010). This microbial activity causes a reducing, alkaline environment where mobility and thereby the concentration of As increases. The effect of microbial activity and organic matter on As mobilization in groundwater in the upstream of Ganges River basin, The Terai Region of Nepal has been shown recently by Yadav et al. (2015).

Arsenic is a member of the carcinogenic heavy metals and the intensity of studies to assess its presence is an outcome of this basic fact. That is why researchers are trying to locate its presence—favoring conditions for elevated concentrations and methods of fighting against such. When the processes and conditions are favorable, elevated As levels are on the agenda. Karstic formations (sinkhole, cave, and conduit), carbonate bedrocks and hydraulic fracturing are such favorable conditions that affect mobility of As in aquifers and increase the interaction between surface and ground water through fractures (Lang et al. 2006; Wang and Luo 2001; Sophocleous 2002). Due to such increased interaction, As spreads to different areas together with these water resources, thereby affecting large masses.

Drainage/irrigation channels affect biogeochemical processes of As distribution. Contamination of soils/sediments with As varies significantly with such channels and depth of the water table. The main reason for this is the passage of oxygen-rich waters from the irrigation channels towards the drainage channels underground, enhancing adsorption of As onto the sediments (Stute et al. 2007; Guo et al. 2011). Guo et al. (2011) also concluded that oxic conditions increase as the distance from channels decreases, giving way to a decrease in total As and As(III) concentration.

Fluctuation in groundwater level due to irrigation practices has also a great effect on As concentrations in groundwater. Shallow aquifers recharged by surface water are generally rich in As during dry/irrigation seasons (Cheng et al. 2005; Oinam et al. 2011). By installing samplers along groundwater flow in Hetao basin-Inner Mongolia and monitoring As presence, Guo et al. (2013) suggested that an increase in As concentrations with high groundwater levels during irrigation seasons (in summer between April and September and in winter in November) can trigger a release of As from sediments to groundwater due to dissolution of iron oxyhydroxides and/or reductive desorption of As(V) in environments favoring reduction. On the other hand, Chauhan et al.'s (2009) study carried out at the Ballia District-UP-India demonstrated no important seasonal difference between groundwater As concentrations in three different seasons (summer, monsoon and winter). Similar results were also obtained by Cheng et al. (2005) at Araihazar, Bangladesh, in a study repeated for 3 consequent years. The majority of scientific studies on the effect of seasonal variations, especially rainy seasons, showed that As concentrations varied according to seasons. The latter two studies mentioned conclude the opposite. It is evident that rainwater induced seepage is not the only factor affecting infiltration of As from sediments to groundwater. Thus, future studies should elaborate more on other probable factors in these areas.

In addition to effect of drainage/irrigation channels and seasonal variations on As concentration in groundwater, depth is an equally important parameter to consider. British Geological Survey (2001) and Chauhan et al. (2009) suggested utilization of potable water from deep wells, since they contain less As concentrations, when compared with shallow wells. According to Yadav et al. (2015), the most important reason forwarded is contamination of shallow groundwater with agrochemical wastes. However, we must bear in mind the fact that lowering of the groundwater table by excessive extraction can cause As infiltration into deeper aquifers (Van Geen et al. 2003; Rahaman et al. 2013).

Grain size of sediments is also another factor which affects the concentration of As in aqueous environments. Yadav et al. (2015) showed that as the grain size decreases, As concentration increases in sediments, since fine grained materials have larger surface areas and thus a high rate of As adsorption. As a result, concentration of As in groundwater can change depending on the grain size of the sediments. Studies on surface waters and sediments in Hongfeng and Baihua reservoirs in China showed that heavy metals tend to be higher in sediments than in water, due to adsorption effect of heavy metals on the suspended solids and settling of sediments on the lake bed (Wu et al. 2014).

4 Transfer from Soil-Water to Plants

In discussing the uptake mechanism of arsenic (As) by plants when it is present in soil in appreciable concentrations, some parameters that need to be taken into consideration are pH, organic matter, redox potential, clay content, water and nutrient availability and microbial activity. These parameters also influence translocation of As to different plant organs, as well as its forms and amounts in soil (Kabata-Pendias and Pendias 2001; Inacio et al. 2013). Uptake of As by plants decreases as Fe, Al, Mn oxides/hydroxides and organic matter rich clays in soil increase. The reason for this is that in the presence of such media, As is adsorbed by the soil particles (Yong and Mulligan 2004; Huang et al. 2006; Hsu et al. 2012). It has been shown by Warren et al. (2003) that As concentration in lettuce, radish, and cauliflower decreased by 22% due to addition of ferrous sulfate in soil. In addition, in the same study bioavailability of As was higher in sandy soils compared to clay soils. This is because high sorption capacity of clay causes a decrease in bioavailability of As. For this reason, As phytotoxicity is more likely to occur relatively less in clay soils than in coarse textured soils. In addition, the uptake and translocation of metals in plants is affected by season, age of plant, nutritional status, any disorders, stage of growth and rhizosphere interplay with micro-organisms (Kříbek et al. 2014).

Ramirez-Andreotta et al. (2013) explained the difference of As concentrations in plants using the bioaccumulation factor (BCF), which is the ratio of metal concentration in the plant to the metal concentration in soil. Plants with BCF $\geq$ 1 are classified as hyperaccumulators (Vithanage et al. 2011) and these kinds of plants are used for phytoremediation.

However, knowing the As concentration of soil itself is not enough to be able to characterize As uptake by plants; bioavailable species of As in soil should also be known (Smith et al. 2008). This is so because only some forms of heavy metals, which are soluble in the soil solution, are bioavailable to plants (Chojnacka et al. 2005).

Moreover, if the As in soil is in the form that renders it possible to be up taken by plants, it will be possible for As to be accumulated at high levels. Bergqvist et al. (2014) claimed the same and said that the opposite is also true if the form of As and thus its bioavailability is lower. The same study also suggested that low to medium As pollution in soils leads to the predomination of arsenite (AsIII) in carrot, lettuce and spinach; while phytotoxic As concentrations in soils result in arsenate (AsV) predomination. This information regarding the presence of arsenite and arsenate in plants is also stressed by Smith and Steinmaus (2009). As a result, it is said that high bioavailable As and a soil with As concentration below phytotoxic levels might enhance the production of vegetables with As-rich edible parts. According to Dixon (1997) arsenate is taken up by plants in oxidizing environments via phosphate transfer system due to the similarity of the arsenate ion with the phosphate ion. This is possible, because phosphate and arsenate ions are similar and exchange of these ions can take place under convenient conditions. For instance, the phosphate in plants can change place with arsenate and thus the plant becomes contaminated with As. Similar bio-transfer of As to human red blood cells occurs in the presence of phosphate. Especially in carcinogen cells, the phosphate changes place by As and As accumulates in the cells at greater rates. Besides inorganic As species, organic As species exist in plants too, which are produced as a result of methylation of inorganic As (Raab et al. 2007). However, other authors like Lomax et al. (2012), claimed that plants do not have any methylation capacity and the reason of organic As existence in plants is microorganisms. The methylation process is also observed in animal tissues for the transformation of inorganic As into organic form. For instance, Centeno et al. (2002) mentioned that although As can be accumulated in animal tissues, the accumulated As in animal tissues can be partly transformed into organic As through biomethylation processes.

5 Arsenic in Plants

Dietary intake of heavy metals poses risks to both human and animal health. Although wide range of people is exposed to arsenic (As) through the consumption of As containing water; foodstuffs (i.e., crops, vegetables) are also a significant route of exposure for remarkable amount of people. Even though diet plays a significant role in human As exposure, there was lack of data on inorganic As (inAs) in foodstuffs in the literature until a decade ago. According to EFSA (2009) almost 98% of published studies in 15 European countries considered total As in various foodstuffs without differentiating organic and inorganic As species. Since organic As species are less toxic to the human body when compared with inorganic species, future studies should concentrate on such studies, especially on bioavailability and toxicity of As depending on its various species. Thereby, overestimation of health risks will be avoided.

The Joint FAO/WHO Expert Committee on Food Additives set a provisional tolerable daily intake (PTWI) of inAs of 2.1 μg/kg body weight per day in 1983 and a provisional weekly tolerable intake (PWTI) of inAs level of 15 μg/kg body weight per week in 1988. However, the European Food Safety Authority (EFSA) Panel of contaminants in the food chain (2009) concluded that these weekly and daily tolerable limit values set by the Joint FAO/WHO Expert Committee on Food Additives are not valid, since inorganic As leads to lung, urinary bladder and skin cancer even at lower concentrations than tolerable limits set.

Most of the studies carried out were related to rice and rice products, followed by the studies related to fish and seafood, seaweed and algae and beverages especially apple juice consumed mostly by children. However, in this review it was dealt with some tuberous, leafy and fruity vegetables and pulses. Importance was attributed to include only results with standard deviations and so some of the researches were excluded due to the absence of standard deviation values. Unfortunately, some of the vegetables were studied once in recent researches and it was impossible to compare them with other studies On the other hand, it is observed that most of the studies were carried out by researchers in West Bengal-India, mainly because of high occurrence of contamination.

Samal et al. (2011) carried out some studies in West Bengal-India, on various crops which were irrigated by As enriched groundwater and observed considerable variations in the amount of As accumulation in different vegetables. Underground tuberous vegetables (such as arum, radish and potato) contained the highest As concentrations (780 ± 243 μg kg^{-1}, 674 ± 211 μg kg^{-1}, 291 ± 176 μg kg^{-1}) while leafy vegetables (cabbage, amaranthus, spinach) contained the second highest As concentrations (315 ± 69.7 μg kg^{-1}, 265 ± 158 μg kg^{-1}, 270 ± 182 μg kg^{-1}). However, unexpectedly cabbage within leafy vegetables showed higher As concentration than potato.

The third highest group was found to be fruity and fleshy vegetables (bitter gourd, brinjal, tomato, etc.) (262 ± 133 μg kg^{-1}, 217 ± 80 μg kg^{-1}, 84.4 ± 48.5 μg kg^{-1}) and lastly the fourth highest group was pulses (lentil, pea) (24.7 ± 16.7 μg kg^{-1}, 69.2 ± 22.9 μg kg^{-1}). On the other hand, the study at the same time showed that crops which require more water for their growth generally had higher As concentrations than crops which need less water. This finding illustrates that using of As enriched groundwater for irrigation purposes can lead to transfer of As through the water-soil-crop-food chain.

Similarly, Biswas et al. (2012) analyzed some leafy, fruity, tuberous and pulses vegetables grown in As contaminated areas in West Bengal-India and compared those collected from market basket. Thirty-two types of vegetables and seven types of pulses were collected from the agricultural areas and eighteen products (vegetables and pulses) were collected from the market. Among all cultivated vegetables tested in this study, pea and lentil from the pulses family showed the highest As concentrations with 1300 ± 480 μg/kg and 1120 ± 144 μg/kg respectively. These are the highest values of As measured in all vegetables listed in this review. These values are also exceeding the WHO-recommended permissible limit (1000 μg/kg) for foodstuffs. On the other hand, among roots and tubers, arum tuber—as expected—showed the highest concentration 558 ± 73 μg/kg and onion bulb showed the lowest with 187 ± 77 μg/kg. Interestingly, spinach (910 ± 259 μg/kg) had a higher concentration than tuberous vegetables. Except tomato (551 ± 262 μg/kg) and bitter gourd (529 ± 44 μg/kg), fruity vegetables showed again the lowest concentrations. The total As value of vegetables taken from the market appeared to be lower than the level observed in field vegetables. It was reported that vegetables obtained from the market were imported from nonlocal areas. Thus, it can be concluded that only field surveys or only market surveys are not adequate to give actual photograph of As contamination of foodstuffs. Results of this study showed the highest concentrations almost for each kind of vegetables among all other results taken from different studies. Only the concentrations of As in arum and radish were found higher by Samal, although they both tested vegetables cultivated in Nadia District, West Bengal.

In general, Tables 1, 2, 3 and 4, it can be concluded that tuberous vegetables generally have the highest As concentration followed by leafy vegetables, fruity vegetables and pulses. However, the rates of accumulation change considerably depending on effective parameters enhancing uptake. As seen in the study carried out by Biswas et al. (2012), lentil, pea and spinach showed extremely higher concentrations than tuberous vegetable types. Biswas' study is an exception which prevents any generalization. This proves that As concentrations in soil and irrigation water, coupled with environmental factors, have great impact on concentration of As in vegetables. Only those vegetables cultivated under similar conditions can be compared to each other. Transfer factors which consider As concentrations in soil and water should be compared for vegetables cultivated in different agricultural fields. Arum, radish and potato show highest accumulation capabilities with regard to As in tuberous vegetables. Cabbage and amaranth in leafy vegetables have almost similar accumulation capacity in the same study areas.

Table 1 Mean concentrations of total As in edible parts of various tuberous vegetables

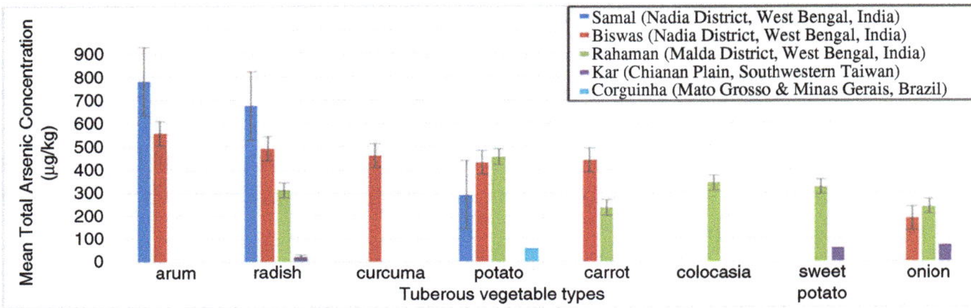

Foodstuff	Type	Concentration (total mean As) (µg/kg)	Origin	Methodology	Reference
Arum tuber	Roots and tubers	780 ± 243 558 ± 73	Nadia District, West Bengal, India Nadia District, West Bengal, India	FI-HG-AAS HG-AAS	Samal et al. (2011) Biswas et al. (2012)
Radish	Roots and tubers	674 ± 211 491 ± 122 312 ± 7 21.5 ± 3.64	Nadia District, West Bengal, India Nadia District, West Bengal, India Malda District, West Bengal, India Chianan Plain, Southwestern Taiwan	FI-HG-AAS HG-AAS HG-AAS GFAAS	Samal et al. (2011) Biswas et al. (2012) Rahaman et al. (2013) Kar et al. (2013)
Curcuma	Roots and tubers	461 ± 128	Nadia District, West Bengal, India	HG-AAS	Biswas et al. (2012)
Potato	Roots and tubers	456 ± 7 431 ± 55 291 ± 176 59 ± 11.58	Malda District, West Bengal, India Nadia District, West Bengal, India Nadia District, West Bengal, India Mato Grosso and Minas Gerais, Brazil	HG-AAS HG-AAS FI-HG-AAS GF-AAS	Rahaman et al. (2013) Biswas et al. (2012) Samal et al. (2011) Corguinha et al. (2015)
Carrot	Roots and tubers	441 ± 53 235 ± 4	Nadia District, West Bengal, India Malda District, West Bengal, India	HG-AAS HG-AAS	Biswas et al. (2012) Rahaman et al. (2013)

(continued)

Table 1 (continued)

Foodstuff	Type	Concentration (total mean As) (μg/kg)	Origin	Methodology	Reference
Colocasia	Roots and tubers	342 ± 7	Malda District, West Bengal, India	HG-AAS	Rahaman et al. (2013)
Sweet potato	Roots and tubers	324 ± 9 58.7 ± 8.17	Malda District, West Bengal, India Chianan Plain, Southwestern Taiwan	HG-AAS GFAAS	Rahaman et al. (2013) Kar et al. (2013)
Onion	Roots and tubers	238 ± 6 187 ± 47 72.2 ± 3.06	Malda District, West Bengal, India Nadia District, West Bengal, India Chianan Plain, Southwestern Taiwan	HG-AAS HG-AAS GFAAS	Rahaman et al. (2013) Biswas et al. (2012) Kar et al. (2013)

In addition, many authors demonstrated that different parts of plants (roots, stems, leaves, fruits) accumulate different amounts of As in their tissues. The descending order of this deposition is illustrated to be in the order root—stem and leaves—fruits (Kar et al. 2013; Dahal et al. 2008, Roychowdhury et al. 2005). It means that, As is taken up by roots from water or soil and translocated to aerial organs in decreasing order from stems to leaves and finally fruits. From the test results, this order of As deposition was observed. For instance, for peas roots appeared to contain 54.3 ± 5.82 μg/kg, stems 24.1 ± 6.18 μg/kg and leaves and edible parts 21.7 ± 6.32 μg/kg. Again, maize roots contained 92.5 ± 8.2 μg/kg, stems 32.5 ± 3.64 μg/kg and leaves and edible parts 19.5 ± 2.68 μg/kg. For rice also, which is one of the plants accumulating comparatively higher As levels, roots contained 182 ± 47.2 μg/kg, stems 47.3 ± 12.5 μg/kg and leaves and edible parts 26.7 ± 39.5 μg/kg, respectively. These results which can be generalized indicating such an order is best observed in the studies of Kar et al. (2013) and Fig. 2 displays these results in a modified version.

6 Arsenic in Rice and Effect of Processing

Inorganic arsenic (inAs) content in a food varies according to food sub-types and samples. FDA (2012) demonstrated for rice-based cereals that three different samples had different concentrations of 16%, 61% and 87% inAs respectively. On the other hand, for brown rice of varying types (jasmine, long-grain, short-grain, sticky) concentrations ranged from 26 (Zavala et al. 2008) to 95% (Huang et al.

Table 2 Mean concentrations of total As in edible parts of various leafy vegetables

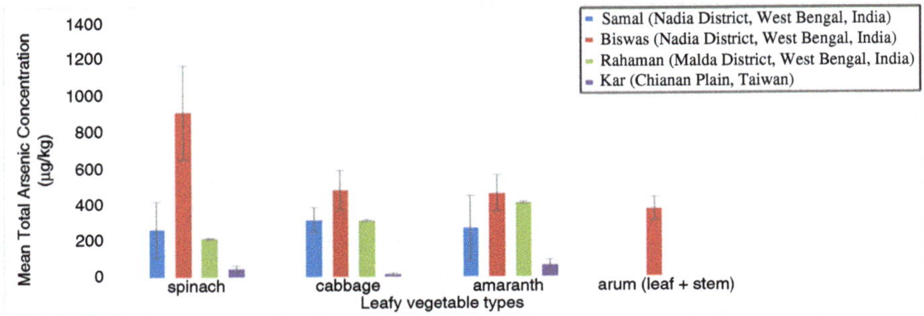

Foodstuff	Type	Concentration (total mean As) (µg/kg)	Origin	Methodology	Reference
Spinach	Leafy vegetable	910 ± 259 265 ± 158 213 ± 3 46.2 ± 16.8	Nadia District, West Bengal, India Nadia District, West Bengal, India Malda District, West Bengal, India Chianan Plain, Southwestern Taiwan	HG-AAS FI-HG-AAS HG-AAS GFAAS	Biswas et al. (2012) Samal et al. (2011) Rahaman et al. (2013) Kar et al. (2013)
Cabbage	Leafy vegetable	482 ± 111 315 ± 69.7 311 ± 5 12.6 ± 6.72	Nadia District, West Bengal, India Nadia District, West Bengal, India Malda District, West Bengal, India Chianan Plain, Southwestern Taiwan	HG-AAS FI-HG-AAS HG-AAS GFAAS	Biswas et al. (2012) Samal et al. (2011) Rahaman et al. (2013) Kar et al. (2013)
Amaranth	Leafy vegetable	462 ± 103 411 ± 7 270 ± 182 63.5 ± 31.5	Nadia District, West Bengal, India Malda District, West Bengal, India Nadia District, West Bengal, India Chianan Plain, Southwestern Taiwan	HG-AAS HG-AAS FI-HG-AAS GFAAS	Biswas et al. (2012) Rahaman et al. (2013) Samal et al. (2011) Kar et al. (2013)
Arum (leaf + stem)	Leafy vegetable	373 ± 68	Nadia District, West Bengal, India	HG-AAS	Biswas et al. (2012)

Table 3 Mean concentrations of total As in edible parts of various fruity vegetables

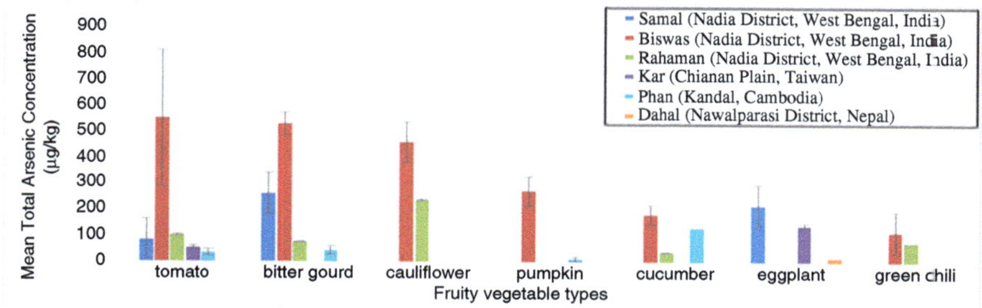

Foodstuff	Type	Concentration (total mean As) (µg/kg)	Origin	Methodology	Reference
Tomatoes	Fruity vegetable	551 ± 262 103 ± 3 84.4 ± 48.5 53.4 ± 7.93 35 ± 13	Nadia District, West Bengal, India Malda District, West Bengal, India Nadia District, West Bengal, India Chianan Plain, Southwestern Taiwan Kandal, Cambodia	HG-AAS HG-AAS FI-HG-AAS GFAAS ICP-MS	Biswas et al. (2012) Rahaman et al. (2013) Samal et al. (2011) Kar et al. (2013) Phan et al. (2013)
Bitter gourd	Fruity vegetable	529 ± 44 262 ± 133 76 ± 1 43 ± 17	Nadia District, West Bengal, India Nadia District, West Bengal, India Malda District, West Bengal, India Kandal, Cambodia	HG-AAS FI-HG-AAS HG-AAS ICP-MS	Biswas et al. (2012) Samal et al. (2011) Rahaman et al. (2013) Phan et al. (2013)
Cauliflower	Fruity vegetable	459 ± 79 237 ± 3	Nadia District, West Bengal, India Malda District, West Bengal, India	HG-AAS HG-AAS	Biswas et al. (2012) Rahaman et al. (2013)
Pumpkin	Fruity vegetable	271 ± 57 11 ± 2	Nadia District, West Bengal, India Kandal, Cambodia	HG-AAS ICP-MS	Biswas et al. (2012) Phan et al. (2013)
Cucumber					

(continued)

Table 3 (continued)

Foodstuff	Type	Concentration (total mean As) (µg/kg)	Origin	Methodology	Reference
	Fruity	181 ± 38 131 ± 9 37 ± 2	Nadia District, West Bengal, India Kandal, Cambodia Malda District, West Bengal, India	HG-AAS ICP-MS HG-AAS	Biswas et al. (2012) Phan et al. (2013) Rahaman et al. (2013)
Eggplant	Fruity vegetable	217 ± 80 140 ± 10 16.9 ± 5.38	Nadia District, West Bengal, India Chianan Plain, Southwestern Taiwan Nawalparasi District, Nepal	FI-HG-AAS GFAAS HG-AAS	Samal et al. (2011) Kar et al. (2013) Dahal et al. (2008)
Green chili	Fruity vegetable	114 ± 81 76 ± 1	Nadia District, West Bengal, India Malda District, West Bengal, India	HG-AAS HG-AAS	Biswas et al. (2012) Rahaman et al. (2013)

2012). Besides the type of product, processing and preparation of food, washing and cooking methods are other factors that lead to a change in As concentration.

The effect of preparation and cooking processes on As retention in rice is well reported in literature. Lynch et al. (2014) suggested that rice and rice products such as bran and rice flours have higher inAs levels than polished rice. Similarly, Rasmussen et al. (2013) carried out some tests on white and brown rice which had same origin and saw that white rice has a lower total level of inAs (160 µg/kg) than brown (400 µg/kg) rice. The reason of this is that as the rice is polished to get whiter, bran which has a higher As concentration, is removed from the surface of rice, resulting in low As concentrations (EFSA 2009). On the other hand, Phan et al. (2013) and Mihucz et al. (2007) suggested based on their studies that, rinsing rice with As free water before cooking might eliminate As from As laden rice, resulting in less As in cooked rice than that in uncooked rice. However, in this case essential trace elements (such as Cu, Mn and Zn) enhancing growth and human health, might also be removed along with As (Mihucz et al. 2010). In contrast, it should be born in mind that As concentration in food can be increased by washing it with As bearing water. Availability of especially inorganic As in changing concentrations in water to be used in cooking will change the As content of the food. For example, As bearing water can alter As concentration in rice after cooking due to contaminated water taken up by rice (Signes et al. 2008; Hossain et al. 2012). On the other hand,

Table 4 Mean concentrations of total As in edible parts of various pulses

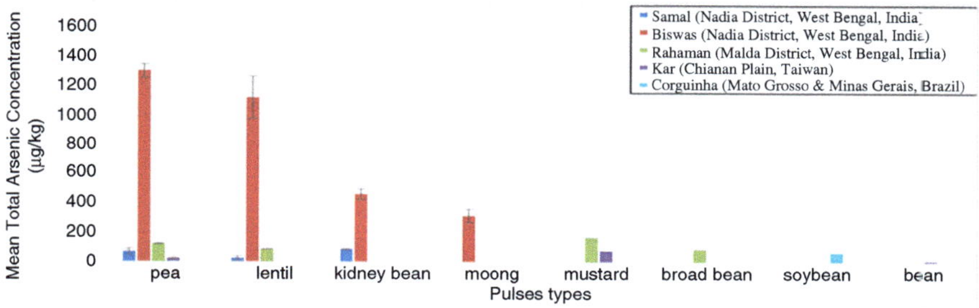

Foodstuff	Type	Concentration (total mean As) (μg/kg)	Origin	Methodology	Reference
Pea	Pulses	1300 ± 48 123 ± 3 69.2 ± 22.9 21.7 ± 6.32	Nadia District, West Bengal, India Malda District, West Bengal, India Nadia District, West Bengal, India Chianan Plain, Southwestern Taiwan	HG-AAS HG-AAS FI-HG-AAS GFAAS	Biswas et al. (2012) Rahaman et al. (2013) Samal et al. (2011) Kar et al. (2013)
Lentil	Pulses	1120 ± 144 89 ± 1 24.7 ± 16.7	Nadia District, West Bengal, India Malda District, West Bengal, India Nadia District, West Bengal, India	HG-AAS HG-AAS FI-HG-AAS	Biswas et al. (2012) Rahaman et al. (2013) Samal et al. (2011)
Kidney bean	Pulses	462 ± 36 85 ± 3.43	Nadia District, West Bengal, India Nadia District, West Bengal, India	HG-AAS FI-HG-AAS	Biswas et al. (2012) Samal et al. (2011)
Moong	Pulses	314 ± 47	Nadia District, West Bengal, India	HG-AAS	Biswas et al. (2012)
Mustard	Pulses	168 ± 5 75.8 ± 22.9	Malda District, West Bengal, India Chianan Plain, Southwestern Taiwan	HG-AAS GFAAS	Rahaman et al. (2013) Kar et al. (2013)
Broad bean	Pulses	88 ± 1	Malda District, West Bengal, India	HG-AAS	Rahaman et al. (2013)
Soybean	Pulses	65 ± 22	Mato Grosso and Minas Gerais, Brazil	GF-AAS	Corguinha et al. (2015)
Bean	Pulses	9.15 ± 1.44	Chianan Plain, Southwestern Taiwan	GFAAS	Kar et al. (2013)

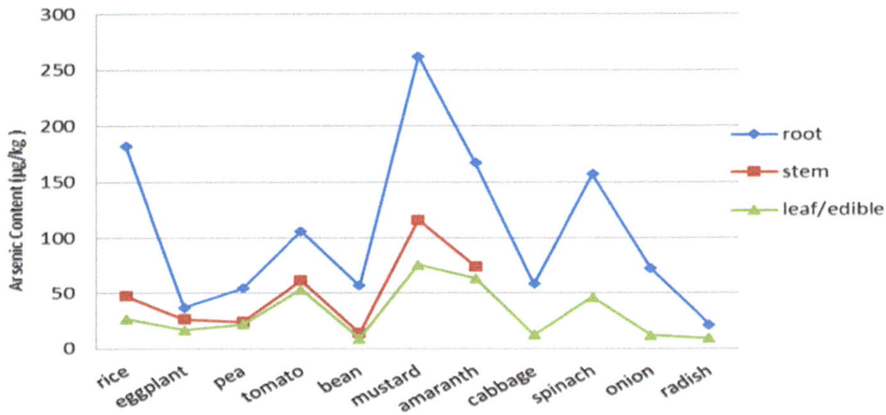

Fig. 2 Variation of As content (μg/kg of fresh weight) in three different parts (root, stem, leaf/ edible part) of vegetables (modified from Kar et al. 2013)

Lynch et al. (2014) reviewed based on their data that cooking rice in a high volume of water reduces the total inAs by up to 20% compared to raw samples as well as samples cooked in a small volume of water.

Processed food is also a significant member of the food chain through which As is taken up by the human body. Sugar et al. (2013) emphasizes this in his study and claims that the As contamination does not only come from the As concentration of the water used in irrigation of the vegetable in question but also from the water used for processing. Sugar et al. (2013) demonstrated this result in the form of a linear relationship between As concentration and amount of water used for processing.

Another study again analysing As content versus water used for food processing revealed similar results. Pinto beans and pasta soup taking up high quantities of water displayed highest As concentrations, whereas tortilla taking up a lower water content displayed lower As concentrations (Del Razo et al. 2002). On the other hand, heating parameters such as period and temperature have an impact on the As concentration of food, since as the water vaporized, the weight of the food is decreased causing higher As concentration in cooked products. Davis et al. (2012) tried to attract attention to rice-related danger for children and suggested that rice consumption is a significant source of As exposure for children. Infants and children are exposed to As 2–3 times more than adults, consuming rice based products such as biscuits, crackers, pasta, noodles, pudding, whole grained rice etc. during their childhood (Da Sacco and Masotti 2012).

7 Conclusion

In conclusion, arsenic (As) is one of the most important toxic and carcinogenic metalloids, which can cause serious health risks to living organisms, exposed through soil, water, air and plants. It can easily mobilize from its source to primary receiving environments (soil, water and air) by a series of pathways causing widespread contamination. The risks in question do not only threaten adults, but also infants and children, due to high As concentrations displayed by rice and rice-based products.

Researchers have concentrated a lot on the mobility and transfer mechanism of As in aqueous, airborne and soil environments. These transfer mechanisms are shown to be dependent on bioavailability of inorganic and organic As, while uptake occurs through favourable conditions of certain parameters such as pH, organic matter, redox potential, clay content, water and nutrient availability and microbial activity. High As concentrations on crops and vegetables were especially due to irrigation with As-enriched waters. It was also shown that As uptake by plants decreases in soils with increasing Fe, Al, Mn oxides/hydroxides and organic matter rich clay, since As is adsorbed by soil particles when these oxides/hydroxides are present. On the other hand, desorption of As from surface of hydrous metal oxides and clay minerals increases under alkaline ($pH > 8$) and oxidative conditions, causing increased As availability in groundwater which can subsequently be taken up by plant roots.

Concentrations of As in each plant species changes depending on its origin (agricultural area, presence/absence of smelters/mines/metallurgical zones), species and contamination rate of agricultural field and nature of irrigation water. Studies also revealed that As accumulation also depends on the different parts of plants. Although not always valid, it can be said in general that tuberous vegetables accumulate the highest As levels and the concentration gradually decreases respectively for leafy vegetables, fruity vegetables and pulses. It must also be born in mind that processed food may also be contaminated depending on washing and cooking methods as well as on the method of processing and rate of absorbance of As-enriched processing water. The majority of researches on As concentration in foodstuffs mostly concentrated on the presence of total As, without giving details on the nature (inorganic, organic) and species of As. More in-depth studies concentrating on separate concentrations of different As species will not only add to scientific database but will also enhance the understanding of related health risks due to their different toxicity effects.

It is not only important to find and analyse the presence and effects of As in soil/air/water-plant-animal-human chain, but it is equally important to continuously monitor As presence. Following the assessment of its amount and its distribution, gaining control and applying remediation measures should continue. Even though it is impossible to get rid of natural As resources, at least anthropogenic As resources such as As containing pesticides, herbicides, chemicals, etc. can be forbidden by the authorities.

8 Summary

This review deals with exposure pathways of arsenic, as well as its transfer and uptake processes from its source to the human body. It is a proven fact that uptake of inorganic As for a long period can lead to chronic As poisoning and a variety of adverse health effects such as skin, lung and bladder cancer, in addition to cardiovascular diseases, diabetes and gastrointestinal symptoms. This risk makes it necessary to continuously search, analyze, monitor and control As contamination of each and every food chain medium (soil, water, plant, animal and human). The danger is not only imposed on adults, but also on infants and children through high consumption of rice-based foodstuffs.

Arsenic exposure occurs primarily from consumption of potable water containing high concentrations of inorganic As. Secondary and almost equally important exposure can be summarized as follows: consumption of crops cultivated in As contaminated (natural or anthropogenic) agricultural fields, consumption of crops cultivated in air-contaminated soils, and consumption of crops irrigated by As-enriched waters. Studies carried out on processed food especially washed and cooked with As-enriched waters demonstrated As contamination, also depending on the water absorption capacity of food type. Thus, consumption of this processed food is also an important As exposure.

The geographical location of the agricultural area is also very important. Detailed study on these areas generally consider the As threat twofold: is it natural or is it anthropogenic. When there are smelters, metallurgical activities, mining activities and similar activities in the surrounding areas introducing chemicals (As and others) to air, soil and water, there is already a risk. Numerous studies demonstrated an increase in the contamination levels of soils and waters in these areas, but under certain circumstances (airborne As and favorable wind speed and direction) accumulation can be inversely proportional with the distance from the contaminated zone. Therefore, continuous monitoring and control measures in such areas are of maximum priority.

In this review, light is also shed on the transfer mechanism of As through the food chain and the parameters that enhance mobility of As in the environment. Since high levels of As accumulation in edible parts of vegetables and crops will be transferred to the human body under convenient conditions, As accumulation in different vegetables and crops and in different parts of these vegetables and crops must also be understood. Numerous relevant studies illustrated that accumulation in descending order in different parts of the vegetables and crops is as follows: roots—stems—leaves and edible parts. Such comparative information, amounts of accumulation in different vegetables and crops as well as in different parts of these vegetables and crops will determine the level of risk imposed on human body. These data will orientate authorities to take necessary precautions to protect the health of people, especially infants and children.

Acknowledgments This work was supported by Cyprus International University Biotechnology Research Center and Cancer Research Foundation in North Cyprus.

References

Aksoy N, Şimşek C, Gunduz O (2009) Groundwater contamination mechanism in geothermal field, a case study of Balcova, Turkey. J Contam Hydrol 103:13–28

Ali MHH, Al-Qahtani KM (2012) Assessment of some heavy metals in vegetables, cereals and fruits in Saudi Arabian markets. Egypt J Aquat Res 38:31–37

Alsina M, Saratovsky I, Gaillard JF, Pasten PA (2007) As speciation in solid phases of geothermal fields. In: Barnett MO, Kent DB (eds) Adsorption of metals by geomedia II: variables, mechanisms and model applications. Elsevier, Amsterdam, pp 17–440

Bergqvist C, Herbert R, Persson I, Greger M (2014) Plants influence on As availability and speciation in the rhizosphere, roots and shoots of three different vegetables. Environ Pollut 184:540–546

Bhattacharya P, Jacks G, Jana J, Sracek A, Gustafsson JP, Chatterjee D (2001) Geochemistry of the Holocene alluvial sediments of Bengal Delta Plain from West Bengal, India: implications on As contamination in groundwater. In: Jacks G, Bhattacharya P, Khan AA (eds) Groundwater As contamination in the Bengal Delta Plain of Bangladesh . KTH Special Publication, Dhaka, pp 21–40TRITA-AMI Report 3084

Bhattacharya P, Claesson M, Bundschuh OS, Fagerberg J, Jacks G, Martin RA, Storniolo AR, Thir JM (2006) Distribution and mobility of As in the Rio Dulce alluvial aquifers in Santiago del Estero Province, Argentina. Sci Total Environ 358:97–120

Bhattacharya P, Samal AC, Majumdar J, Santra SC (2010) As contamination in rice, wheat, pulses and vegetables: a study in an As affected area of West Bengal. India. Water Air Soil Pollut 213:3–13

Biswas A, Biswas S, Santra SC (2012) Risk from winter vegetables and pulses produced in As endemic areas of Nadia District: field study comparison with market basket survey. Bull Environ Contam Toxicol 88:909–914

Bondada BR, Underhill RS, Ma LQ, Davidson MR, Guyoda Y, Duran RS (2007) Spatial distribution, localization and speciation of As in the hyperaccumulating fern (*Pteris vittata* L.). In: Bhattacharya P, Mukherjee AB, Loeppert RH (eds) Arsenic in soil and groundwater environments: trace metals and other contaminants in environment. Elsevier, Amsterdam, pp 299–314

British Geological Survey (BGS) (2001) As contamination of groundwater in Bangladesh. In: Kinniburgh DG, Smedley PL (eds) Final report, BGS technical report WC/00/19. NERC, DPHE, DFID, Dhaka

Bundschuh J, Maity JP, Nath B, Baba A, Gunduz O, Kulp TR, Jean JS, Kar S, Yang HJ, Tseng YJ, Bhattacharya P, Chen CY (2013) Naturally occurring As in terrestrial geothermal systems of Western Anatolia: potential role in contamination of freshwater resources, Turkey. J Hazard Mineral 262:951–959

Centeno J, Mullick FG, Martinez L, Page N, Gibb H, Longfellow D (2002) Pathology related to chronic As exposure. Environ Health Perspect 110:883–886

Chatterjee D, Halder D, Majumder S, Biswas A, Nath B, Bhattacharya P, Bhowmick S, Mukherjee Goswami A, Saha D, Hazra R, Maity PB, Chatterjee D, Mukherjee A, Bundschuh J (2010) Assessment of As exposure from groundwater and rice in Bengal delta region, West Bengal, India. Water Res 44:5803–5812

Chauhan VS, Nickson RT, Chauhan D, Iyengar L, Sankararamakrishnan N (2009) Groundwater geochemistry of Ballia district, Uttar Pradesh, India and mechanism of As release. Chemosphere 75:83–91

Chen TB, Song B, Zheng YM, Huang ZC, Zheng GD, Li YX et al (2006) A survey of As concentrations in vegetables and soils in Beijing and the potential risks to human health. Acta Geograph Sin 61(3):297–310
Cheng Z, Van Geen A, Seddique AA, Ahmed KM (2005) Limited temporal variability of As concentrations in 20 wells monitored for 3 years in Araihazar, Bangladesh. Environ Sci Technol 39:4759–4766
Chojnacka K, Chojnacki A, Górecka H, Górecki H (2005) Bioavailability of heavy metals from polluted soils to plants. Sci Total Environ 337:175–182
Chung JY, Yu SD, Hong YS (2014) Environmental Source of As Exposure. J Prev Med Public Health 47:253–257
Corguinha APB, Souza GA, Gonçalves VC, Carvalho CA, Limaa WEA, Martins FD, Yamanaka CH, Francisco EAB, Guilherme LRG (2015) Assessing arsenic, cadmium, and lead contents in major crops in Brazil for food safety purposes. J Food Compos Anal 37:143–150
Csavina J, Field J, Taylor MP, Gao S, Landazuri A, Betterton EA (2012) A review on the importance of metals and metalloids in atmospheric dust and aerosol from mining operations. Sci Total Environ 43:58–73
Da Sacco L, Masotti A (2012) Children do not like As in their food. J Expo Sci Environ Epidemiol 22(4):424–425
Dahal BM, Fuerhacker M, Mentler A, Karki KB, Shrestha RR, Blum WEH (2008) As contamination of soils and agricultural plants through irrigation water in Nepal. Environ Pollut 155:157–163
Das D, Samanta G, Mandal K, Chowdhury TR, Chanda CR, Chowdhury PP, Basu GK, Chakraborti D (1996) As in ground water in six districts of West Bengal, India. Environ Geochem Health 18:5–15
Davis MA, Mackenzie TA, Cottingham KL, Gilbert-Diamond D, Punshon T, Karagas MR (2012) Rice consumption and urinary As concentrations in US children. Environ Health Perspect 12:1418–1424
Del Razo LM, Garcia Vargas GG, Garcia Salcedo J, Sanmiguel MF, Rivera M, Hernandez MC, Cebrian ME (2002) As levels in cooked food and assessment of adult dietary intake of As in the Region Lagunera, Mexico. Food Chem Toxicol 40:1423–1431
Demirel Z, Yıldırım N (2002) Boron pollution due to geothermal wastewater discharge into the Buyuk Menderes River, Turkey. Int J Environ Pollut 18:602–608
Dixon HBF (1997) The biochemical action of As acids especially as phosphate analogues. Adv Inorg Chem 44:191–227
European Food Safety Authority (EFSA) (2009) Scientific opinion on As in food. Panel on contaminants in the food chain (CONTAM). EFSA J 7(10):1351
Gemici U, Tarcan G (2004) Hydrogeological and hydrochemical feature of the Heybeli Spa, Afyon, Turkey: As and the other contaminants in the thermal waters. Bull Environ Contam Toxicol 72:1107–1114
Goldberg S, Johnston CT (2001) Mechanisms of As adsorption on amorphous oxides evaluated using macroscopic measurements, vibrational spectroscopy and surface complexation modeling. J Colloid Interface Sci 234:204–216
Guo HM, Yang S, Tang X, Li Y, Shen Z (2008) Groundwater geochemistry and its implications for As mobilization in shallow aquifers of the Hetao Basin, Inner Mongolia. Sci Total Environ 393:131–144
Guo HM, Zhang B, Li Y, Berner Z, Tang X, Norra S, Stüben D (2011) Hydrogeological and biochemical constrains of As mobilization in shallow aquifers from the Hetao Basin, Inner Mongolia. Environ Pollut 159:876–883
Guo HM, Zhang Y, Jia YF (2013) Spatial and temporal evolutions of groundwater As approximately along the flow path in the Hetao basin, Inner Mongolia. Chin Sci Bull 58:3070–3079
Hossain MA, Rahman MM, Murrill M, Das B, Roy B, Dey S, Maity D, Chakraborti D (2012) Water consumption patterns and factors contributing to water consumption in As affected population of rural West Bengal, India. Sci Total Environ 463–464:1217–1224

Hsu WM, Hsi HC, Huang YT, Liao CS, Hseu ZY (2012) Partitioning of As in soil-crop systems irrigated using groundwater: a case study of rice paddy soils in southwestern Taiwan. Chemosphere 86:606–613

Huang RQ, Gao SF, Wang WL, Staunton S, Wang G (2006) Soil As availability and the transfer of soil As to crops in suburban areas in Fujian Province, Southeast China. Sci Total Environ 368:531–541

Huang J-H, Fecher P, Ilgen G, Hu KN, Yang J (2012) Speciation of arsenite and arsenate in rice grain—verification of nitric acid based extraction method and mass sample survey. Food Chem 130(2):453–459

Inacio M, Neves O, Pereira V, Ferreira da Silva E (2013) Levels of selected potential harmful elements (PHEs) in soils and vegetables used in diet of the population living in the surroundings of the Estarreja Chemical Complex. Appl Geochem 44:38–44

Islam FS, Gault AG, Boothman C, Polya DA, Charnock JM, Chatterfee D, LIoyd JR (2004) Role of metal reducing bacteria in As release from Bengal delta sediments. Nature 430:68–71

Kabata-Pendias A, Pendias H (2001) Trace elements in soil and plants, 3rd edn. CRC Press, Boca Raton, pp. 225–233

Kapaj S, Peterson H, Liber K, Bhattacharya P (2006) Human health effects from chronic As poisoning—a review. J Environ Sci Health A 41:2399–2428

Kar S, Maity JP, Jean JS, Liu CC, Nath B, Yang HJ, Bundschuh J (2010) As enriched aquifers: occurrences and mobilization of As in groundwater of Ganges Delta Plain, Barasat, West Bengal, India. Appl Geochem 25:1805–1814

Kar S, Das S, Jean JS, Chakraborty S, Chuan-Liu C (2013) As in the water-soil-plant system and the potential health risks in the coastal part of Chianan Plain, Southwestern Taiwan. J Asian Earth Sci 77:295–302

Keil DE, Ritchie JB, McMillin GA (2011) Testing for toxic elements: a focus on As, cadmium, lead, and mercury. Lab Med 42(12):735–742

Kříbek B, Majer V, Knésl I, Nyambe I, Mihaljevič M, Ettler V, Sracek O (2014) Concentrations of As, copper, cobalt, lead and zinc in cassava (*Manihot esculenta* Crantz) growing on uncontaminated and contaminated soils of the Zambian Copperbelt. J African Earth Sci 99:713–723

Lam JC, Sia Su GL (2009) Total As and total mercury concentrations of the waters and janitor fishes (Pterygoplichthys spp.) in the Marikina river, Philippines. J Appl Sci Environ Sanit 4 (1):37–42

Lang YC, Liu CQ, Zhao ZQ, Li SL, Han GL (2006) Geochemistry of surface and ground water in Guiyang, China: water/rock interaction and pollution in a karst hydrological system. Appl Geochem 21:887–903

Liaoa VHC, Chua YJ, Sua YC, Lina PC, Hwangb YH, Liua CW, Liaoa CM, Changa FJ, Yua CW (2011) Assessing the mechanisms controlling the mobilization of As in the As contaminated shallow alluvial aquifer in black foot disease endemic area. J Hazard Mater 197:397–403

Lijzen JPA, Baars AJ, Otte PF, Rikken MGJ, Swartjes FA, Verbruggen EMJ, van Wezel AP (2001) Technical evaluation of the intervention values for soil/sediment and groundwater. RIVM report 711701 023, pp 1–147

Lomax C, Liu WJ, Wu L, Xue K, Xiong J, Zhou J, McGrath SP, Meharg AA, Miller AJ, Zhao FJ (2012) Methylated As species in plants originate from soil microorganisms. New Phytol 193:665–672

Lynch HN, Greenberg GI, Pollock MC, Lewis AS (2014) A comprehensive evaluation of inorganic As in food and considerations for dietary intake analyses. Sci Total Environ 496:299–313

Masscheleyn PH, Delaune RD, Patrick WH (1991) Effects of redox potential and pH on As speciation and solubility in a contaminated soil. Environ Sci Tech 25:1414–1419

Mercado M, Garcia ME, Quintanilla J (2009) Evaluación de los Niveles de Contaminación por Plomo y Arsénico en muestras de Suelos y productos Agrícolas Procedentes de la región cercana al Complejo metalrgico Vinto. Rev Boliv Quím 26(2):101–110

Mihucz VG, Tatar E, Virag I, Zang C, Yao J, Zaray G (2007) As removal from rice by washing and cooking with water. Food Chem 105:1718–1725

Mihucz VG, Silversmit G, Szaloki I, de Samber B, Schoonjans T, Tatar E, Vincze L, Virag I, Yao J, Zaray G (2010) Removal of some elements from washed and cooked rice studied by inductively coupled plasma mass spectrometry synchrotron based confocal micro-X-ray fluorescence. Food Chem 121:290–297

Nickson RT, McArthur JM, Burgess WG, Ahmed KM, Ravenscroft P, Rahman M (1998) As poisoning of Bangladesh groundwater. Nature 395:338

Oinam JD, Ramanathan A, Linda A (2011) A study of As, iron and other dissolved ion variations in groundwater of Bishnupur district, Manipur, India. Environ Earth Sci 62:1183–1195

Ongley LK, Sherman L, Armienta A, Concilio A, Ferguson SC (2007) As in the soils of Zimapán, Mexico. Environ Pollut 145:793–799

Orloff K, Mistry K, Metcalf S (2009) Biomonitoring for environmental exposures to As. J Toxicol Environ Health 12:509–524

Phan K, Sthiannopkao S, Heng S, Phan S, Huoy L, Wong MH, Kim KW (2013) As contamination in the food chain and its risk assessment of populations residing in the Mekong River basin of Cambodia. J Hazard Mater 262:1064–1071

Raab A, Ferreira K, Meharg AA, Feldmann J (2007) Can As phytochelatin complex formation be used as an indicator for toxicity in *Helianthus annuus*? J Exp Bot 58:1333–1338

Rahaman S, Sinha AC, Pati R, Mukhopadhyay D (2013) As contamination: a potential hazard to the affected areas of West Bengal, India. Environ Geochem Health 35:119–132

Ramirez-Andreotta MD, Brusseau ML, Artiola JF, Maier RM (2013) A greenhouse and field-based study to determine the accumulation of As in common homegrown vegetables grown in mining-affected soils. Sci Total Environ 443:299–306

Ramos OE, Rötting TS, French M, Sracek O, Bundschuh J, Quintanilla J, Bhattacharya P (2014) Geochemical processes controlling mobilization of As and trace elements in shallow aquifers and surface waters in the Antequera and Poopo mining regions, Bolivian Altiplano. J Hydrol 518:421–433

Rango T, Vengosh A, Dwyer G, Bianchini G (2013) Mobilization of As and other naturally occurring contaminants in groundwater of the Main Ethiopian Rift aquifers. Water Res 47:5801–5818

Rasmussen RR, Qian Y, Sloth JJ (2013) SPE HG-AAS method for the determination of inorganic As in rice-results from method validation studies and a survey on rice products. Anal Bioanal Chem 405(24):7851–7857

Ravenscroft P, Brammer H, Richards KS (2009) As pollution: a global synthesis. Wiley, New York

Roychowdhury T, Tokunaga H, Uchino T, Ando M (2005) Effect of As contaminated irrigation water on agricultural land soil and plants in West Bengal, India. Chemosphere 58:799–810

Samal AC, Kar S, Bhattacharya P, Santra SC (2011) Human exposure to As through foodstuffs cultivated using As contaminated groundwater in areas of West Bengal, India. J Environ Sci Health 46:1259–1265

Signes A, Mitra K, Burlo F, Carbonell-Barrachina AA (2008) Contribution of water and cooked rice to an estimation of the dietary intake of inorganic As in a rural village of West Bengal, India. Food Addit Contam 25:41–50

Smedley PL, Kinniburgh DG (2002) A review of the source, behavior and distribution of As in natural waters. Appl Geochem 17:517–568

Smith AH, Steinmaus CM (2009) Health effects of As and chromium in drinking water: recent human findings. Annu Rev Public Health 30:107–122

Smith AH, Lingas EO, Rahman M (2000) Contamination of drinking-water by As in Bangladesh: a public health emergency. Bull World Health Organ 78(9):1093–1103

Smith PG, Koch I, Reimer KJ (2008) Uptake, transport and transformation of arsenate in radishes (*Raphanus sativus*). Sci Total Environ 390:188–197

Sophocleous M (2002) Interactions between groundwater and surface water: the state of the science. Hydrgeol J 10:52–67

Sracek O, Armienta MA, Rodriguez R, Villasenor G (2010) Discrimination between diffuse and point sources of As at Zimapan, Hidalgo state, Mexico. J Environ Monit 12:329–337

Stute M, Zheng Y, Schlosser P, Horneman A, Dhar RK, Datta S, Hoque MA, Seddique AA, Shamsuddduha M, Ahmed KM, Van Geen A (2007) Hydrological control of As concentrations in Bangladesh groundwater. Water Resour Res 43:W09417. doi:10.1029/2005wr004499

Sugar E, Tatar E, Zaray G, Mihucz V (2013) Relationship between As content of food and water applied for food processing. Food Chem Toxicol 62:601–608

US Food and Drug Administration (FDA) (2012) Arsenic in rice: full analytical results from rice/rice product sampling—September 2012, 9 pp. http://www.fda.gov/Food/FoodborneIllness Contaminants/Metals/ucm319916.htm

Van Geen A, Zheng Y, Versteeg R, Stute M, Horneman A, Dhar R, Steckler M, Gelman A, Small C, Ahsan H, Graziano JH, Hussain I, Ahmed KM (2003) Spatial variability of As in 6000 tube wells in a 25 km^2 area of Bangladesh. Water Resour Res 39:1440

Vithanage M, Dabrowska BB, Mukherjee AB, Bhattacharya P (2011) As uptake by plants and possible phytoremediation applications: a brief overview. Environ Chem Lett 10:217–224

Wang Y, Luo TMZ (2001) Geostatistical and geochemical analysis of surface water leakage into groundwater on a regional scale: a case study in the Liulin karst system, northwestern China. J Hydrol 246:223–234

Wang B, Liu L, Gao Y, Che J (2009) Improved phytoremediation of oilseed rape (*Brassica napus*) by Trichoderma mutant constructed by restriction enzyme-mediated integration (REMI) in cadmium polluted soil. Chemosphere 74:1400–1403

Warren GP, Alloway BJ, Lepp NW, Singh B, Bochereau FJ, Penny C (2003) Field trials to assess the uptake of As by vegetables from contaminated soils and soil remediation with iron oxides. Sci Total Environ 311(1–3):19–33

Webster JG, Nordstrom DK (2003) Geothermal As. In: Welch AH, Stollenwerk KG (eds) As in ground water: geochemistry and occurrence. Kluwer Academic, Boston, pp 101–126

Wu B, Wang G, Wu J, Fu Q, Liu C (2014) Sources of heavy metals in surface sediments and an ecological risk assessment form two adjacent plateau reservoirs. PLoS One 9(7):e102101

Yadav IC, Devi NL, Mohan D, Shihua Q, Singh S (2014) Assessment of groundwater quality with special reference to arsenic in Nawalparasi district, Nepal using multivariate statistical techniques. Environ Earth Sci 72(1):259–273

Yadav IC, Devi NL, Singh S (2015) Reductive dissolution of iron oxyhydroxides directs groundwater As mobilization in the upstream of Ganges river basin, Nepal. J Geochem Explor 148:150–160

Yan XP, Kerrich R, Hendry MJ (2000) Distribution of As (III), As (V) and total inorganic As in pore waters from a thick till and clay-rich aquitard sequence, Saskatchewan, Canada. Geochim Cosmochim Acta 62:2637–2648

Yong RN, Mulligan CN (2004) Natural attenuation of contaminants in soils. CRC Press, Boca Raton

Zavala YJ, Gerads R, Gorleyok H, Duxbury JM (2008) As in rice: II. As speciation in USA grain and implications for human health. Environ Sci Technol 42(10):3861–3866

Modelling the Release, Transport and Fate of Engineered Nanoparticles in the Aquatic Environment – A Review

Adriaan A. Markus, John R. Parsons, Erwin W.M. Roex, Pim de Voogt, and Remi W.P.M. Laane

Contents

1 Introduction 54
2 Previous Overviews and Critical Publications 58
3 Small-Scale Modelling 60
  3.1 Population Balance Models 60
  3.2 Mass Concentrations 61
  3.3 Dissolution 62
  3.4 Chemical and Physical Reactivity 63
  3.5 DLVO Theory 63
  3.6 Particle Dynamics 65
4 Intermediate Scale: Environmental and Technical Compartments 66
  4.1 Groundwater and Soil 66
  4.2 Surface Waters 68

R.W.P.M. Laane (deceased)

A.A. Markus (✉)
Deltares, P.O.Box 177, Delft 2600 MH, The Netherlands

Earth Surface Science, IBED, University of Amsterdam, Science Park 904, Amsterdam 1098 XH, The Netherlands
e-mail: arjen.markus@deltares.nl

J.R. Parsons
Earth Surface Science, IBED, University of Amsterdam, Science Park 904, Amsterdam 1098 XH, The Netherlands
e-mail: j.r.parsons@uva.nl

E.W.M. Roex
Deltares, P.O.Box 177, Delft 2600 MH, The Netherlands
e-mail: erwin.roex@deltares.nl

P. de Voogt
Aquatic Environmental Ecology, IBED, University of Amsterdam, Science Park 904, Amsterdam 1098 XH, The Netherlands

KWR Watercycle Research Institute, Nieuwegein, The Netherlands
e-mail: w.p.devoogt@uva.nl

P. de Voogt (ed.), *Reviews of Environmental Contamination and Toxicology Volume 243*, Reviews of Environmental Contamination and Toxicology 243,
DOI 10.1007/398_2016_17

4.3 Wastewater Treatment Plants and Other Technical Compartments 69
4.4 Probabilistic Methods 70
5 Large-Scale Modelling 71
5.1 Multimedia Models 71
5.2 Material Flow Analysis and Exposure Modelling 72
5.3 GIS Approaches 74
6 Conclusions and Summary 75
6.1 Data on Use, Production and Emission 76
6.2 Modelling Processes in the Environment 78
6.3 Open Questions and Guidelines 79
A Mathematical Details 81
A.1 Population Balance Theory 81
A.2 DLVO Theory 81
A.3 Transport and Adsorption in Groundwater 82
References 83

1 Introduction

Nanoparticles are a product of nanotechnology. In a broad sense, nanotechnology is the science and technology of manipulating matter at very small scales, down to a few nanometers or even smaller. This leads to nanomaterials, where structures of a few hundred nanometers or less are used for the actual functionality and more specifically to nanoparticles, roughly particles up to 100 nm in size in at least one dimension [see (European Commission 2011) for a refined definition]. Their small size causes them to have specific properties that are useful for widely varying purposes, such as antimicrobial and semiconductor properties or a colour that depends on the particle size. Not only the size and composition characterise nanoparticles, but also the crystalline structure, the form and the coating that is applied for functionalisation are important features (Baumann et al. 2014; Jarvie and King 2010).

Many applications of nanotechnology have been suggested, such as smart medicines that deliver the active ingredients at the desired location in the human body. However, to date most actual applications are found in consumer products like cosmetics, food and food packaging, paints and coatings (Hansen et al. 2016).

With the increasing use of nanoparticles in consumer products and industry the need to understand the possible consequences for human and environmental health is also increasing. This is reflected in the number of publications that concern the possible toxicological and ecotoxicological effects of these relatively new materials but also in the number of publications devoted to the release, transport and fate of nanoparticles in especially the aquatic environment (Peralta-Videa et al. 2011; Bour et al. 2015; Doiron et al. 2012; Farmena et al. 2012; Eduok et al. 2013; Fabrega et al. 2011; Gao et al. 2009). Of the many types of nanoparticles that have been studied, a small number is most commonly described in the literature: nanoparticles based on metal oxide, such as titanium dioxide, zinc oxide and cerium oxide, nanoparticles based on metal, for instance silver and iron, and carbon-based

Table 1 Number of publications reported by Google Scholar that mention nanoparticles of a particular material

Material	Total number of publications	Number of publications also mentioning "environment"
Carbon nanotubes	717,000	305,000
Fullerenes	64,900	39,000
Cerium (di)oxide	75,000	54,000
Copper oxide	1,130,000	390,000
Gold	1,280,000	697,000
Iron (zero-valent)	890,000	411,000
Silica	1,130,000	633,000
Silver	856,000	401,000
Titanium dioxide	179,000	103,000
Zinc oxide	436,000	184,000

nanoparticles, fullerenes and their derivatives, as well as carbon nanotubes. They are also the types most commonly applied in consumer products (Hansen et al. 2016). In Table 1 a summary is given of the number of publications that mention nanoparticles of a certain material.

When these nanomaterials enter the environment, they may have adverse effect on organisms through various mechanisms. The significance of these effects will depend on the actual exposure, which in turn depends on concentration and state of the nanomaterials. Understanding these factors is crucial to the assessment of the risks that nanomaterials present.

Some nanomaterials, like zero-valent iron nanoparticles, might be introduced deliberately to remediate soil contamination by a wide variety of substances (Cundy et al. 2008). In that case it will be necessary to understand how the nanoparticles are transported through the soil and subsequently how they get into contact with these contaminants and how they react with them. Similarly, nanomaterials like titanium dioxide are being investigated for the cleansing of wastewater (see for instance (Mohapatra et al. 2014b)).

An important issue that influences any research into the above questions is the fact that there are currently limited measurement techniques to detect and quantify nanoparticles in such complex matrices as natural waters or soils, although new techniques are being developed (Wagner et al. 2014).

Mathematical modelling techniques can be used to accommodate for this situation. In addition they can be used to analyse scenarios about current and future developments, for example when it comes to release into the environment via the waste stream. Modelling techniques can also help to design the use of nanomaterials for soil remediation or treatment of wastewater.

Another aspect of nanoparticles in the environment where modelling techniques can help is that of understanding and describing the various processes nanoparticles are subjected to, for instance aggregation but also dissolution and chemical transformation. These processes have consequences for the transport of nanoparticles but also for the potential ecotoxicological effects.

In this article publications on the various modelling approaches are reviewed. We focus on publications that describe models for the release, transport and fate of nanoparticles in surface water and soil. In particular, no attempt was made to comprehensively review the literature on modelling toxicity or ecotoxicity of this type of contaminants.

For the review the publications have been divided in three groups to clarify the sort of questions the described techniques can be applied to:

- *Small scale:* Several common theories recur, especially within the literature that deals with laboratory experiments. The DLVO theory, developed by Derjaguin, Landau, Verwey and Overbeek, is commonly used to predict or explain the stability of dispersions of nanoparticles by means of the potential energy caused by van der Waals forces and electrostatic repulsion. For transport of nanoparticles in soil almost all articles use the classical colloid filtration theory. To describe the aggregation of nanoparticles to clusters or adsorption to suspended particulate matter, often population balance theories are used. Here size classes of nanoparticles and aggregates of these particles are considered. An alternative approach is to use the advection-diffusion equation with additional terms to represent these and other processes.

  This type of modelling techniques is especially useful for understanding laboratory experiments, but some can also be applied in the studies into the transport and fate in the environment.
- *Intermediate scale:* The environmental compartments most commonly encountered are: soil and fresh water as well as "technical" compartments like wastewater treatment plants and waste incinerator installations. Surprisingly, no articles were found that consider the modelling of (engineered) nanoparticles in marine or estuarine environments.

  For the description of processes like aggregation and dissolution use can be made of the insights from "small-scale" modelling approaches. In addition, however, it is necessary to know or at least to be able to estimate the release of nanomaterials into the system under study. Here techniques that are generally applied with the large scale in mind can be useful.
- *Large scale:* Articles concerned with the large scale, be it the whole world, a country or a river basin, include: estimation of the world-wide production of nanoparticles, life cycle analysis and multimedia models that consider "generic," fully-mixed, environmental compartments (Mackay et al. 2001).

  Since to date we have no adequate data on the release of nanoparticles into the aquatic or terrestrial environment, we need to rely on analyses of their production, use in consumer products and the use of these products.

To some extent the spatial scale is correlated with the level of detail or the complexity of the modelling approach (see Fig. 1): the larger the spatial scale the less details are included in the modelling with respect to the processes that nanoparticles are subject to, and vice versa.

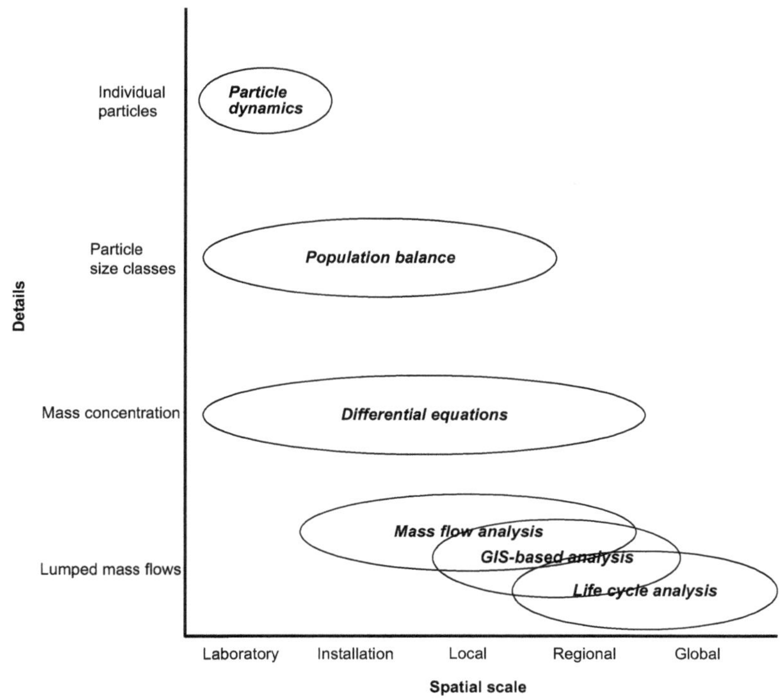

Fig. 1 Schematic classification of the various modelling methods. The vertical axis represents the level of detail these methods provide or require, whereas the horizontal axis represents the spatial scale at which they are typically applied

It is not always possible to strictly distinguish between modelling techniques. Often an actual model will combine several of the techniques described here. For instance, a "global" technique such as life cycle analysis requires some understanding of the behaviour of the contaminants in the environment and a model focusing on the transport of nanoparticles in a river system relies on estimates of emissions obtained perhaps from analyses of production data. The classifications presented here serve as a tool for discussion and interpretation.

In the present article we provide an overview of the modelling approaches that have been published to date. These modelling approaches all have their pros and cons, while none is suitable to answer all questions. Which one to choose depends on the particular questions that need to be answered. Spatial scale and level of detail as well as the environmental compartment of interest determine to a large extent which approach is suitable. Therefore the article concludes with some suggestions as to how to choose the most appropriate model.

2 Previous Overviews and Critical Publications

The literature on nanoparticles has been summarized in many reviews, concentrating on specific aspects. The subject of modelling the release, transport and fate of nanoparticles is no exception. Each group of authors of course has their own emphasis and goal with these reviews. For instance, Hendren et al. (2013a) have reviewed the various approaches to modelling environmental exposure with specific emphasis on the use for risk-based decision making. A particular point of attention is whether the approaches allow for handling uncertainty: the scarcity of experimental and environmental data regarding the fate of nanoparticles make it necessary to explicitly deal with this issue. They plea for the use of methods such as probabilistic modelling and sensitivity analysis and for a closer cooperation between modellers and experimentalists.

The overview by Wagner et al. (2014) is meant to highlight the differences and the similarities between engineered and natural nanoparticles. They describe the range of environmental processes that nanoparticles are subject to and use generally encountered modelling concepts to illustrate these processes.

In a more or less similar way, but with emphasis on the modelling techniques, Dale et al. (2015a) present an introduction to the current state of fate models for nanoparticles. Their message is that while much progress has been made over the past years, existing fate models concentrate on processes such as heteroaggregation, dissolution and sedimentation (see Fig. 2). However, current models cannot account for the influence of coatings or the various environmental conditions, such as pH, temperature and the presence of oxygen and sulfide.

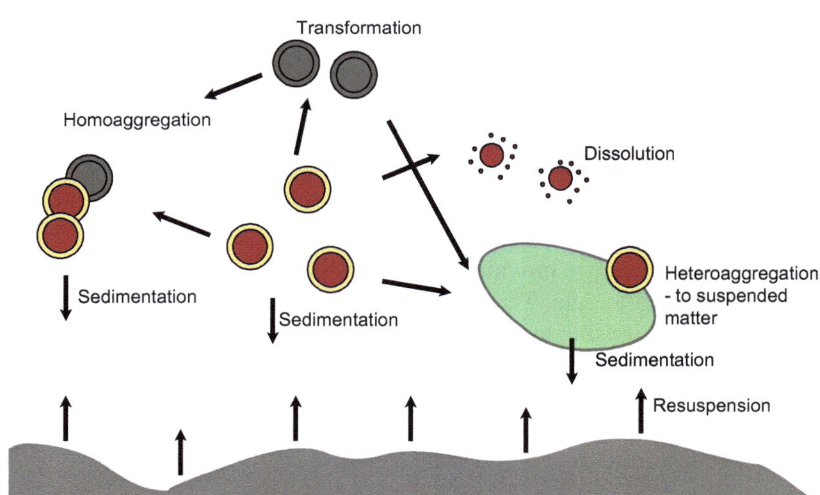

Fig. 2 Sketch of the most important processes that nanoparticles are subject to. [Reproduced from (Markus et al. 2015)]

Praetorius et al. (2014) argue about a particularly common approach to deal with one aspect of the fate of nanoparticles, i.e. the adsorption of nanoparticles to suspended particulate matter (SPM), and the misuse of the concept of partition coefficients in models to describe this phenomenon. They observe that, unlike ordinary, dissolved, contaminants, nanoparticles form an unstable suspension, where there is no thermodynamic equilibrium. The absence of such an equilibrium is the reason partition coefficients cannot be used for describing the adsorption of nanoparticles.

Other aspects of the fate of nanoparticles in the environment have been the subject of critical examination too. Goldberg et al. (2014) conclude after examining the performance of various models for the transport of nanoparticles in saturated soil, that none of the existing models is really satisfactory. Instead of abandoning these models, however, they suggest to investigate the performance of the models separately for retention profiles and breakthrough curves. Furthermore they suggest to critically examine the available data and to make sure that the data are sufficient for calibrating the models.

General guidelines with respect to the use of multimedia models can be found in a publication by Buser et al. (2012). While they do not consider nanoparticles, their advices should hold for any model that employs the multimedia idea (see also Sect. 5.1). The gist of these guidelines is:

- The purpose of the modelling should be clear, that is: what decision-related questions are considered, what substances are relevant in what environmental setting and what are the technical requirements for the modelling.
- The modelling should be reproducible by independent researchers, so that it is necessary to describe what model is used and what input is used, for example, besides emission data also the model parameters. Also document the origin of these data.
- Describe the output of the model and show via a sensitivity analysis what input data have the largest influence on the output. Furthermore the limitations of the model and the limits of the applicability of the results should be clarified. This will help researchers and decision makers to interpret the results.

A comprehensive comparison between experimental studies and modelling studies has been made by Gottschalk et al. (2013), with the conclusion that the two types of research are in fact largely in agreement as to the expected concentrations of engineered nanoparticles in the environment (see also Table 2). They refer to knowledge gaps and scarcity of data with respect to production, application and release, but also to the problem that model results and measurements do not always allow for a comparison due to the differences in particle size and form that are studied.

Table 2 Concentrations of engineered nanoparticles in surface water and soils (treated with sewage sludge and untreated)—results of modelling studies and direct measurements. Adapted from (Gottschalk et al. 2013)

Type of nanoparticle	Water (μg/l)	Soil (μg/kg)
nano-TiO_2 (modelled)	$2 \times 10^{-2} - 10^{1}$	$7 \times 10^{-5} - 2 \times 10^{3}$
nano-TiO_2 (measured)	$3 \times 10^{-1} - 10^{1}$	–
nano-Ag (modelled)	$10^{-5} - 4 \times 10^{-1}$	$8 \times 10^{-5} - 4 \times 10^{-1}$
nano-Ag (measured)	$7 \times 10^{-3} - 3 \times 10^{-3}$	–
nano-ZnO (modelled)	$2 \times 10^{-4} - 10^{-1}$	$4 \times 10^{-1} - 3 \times 10^{2}$
carbon nanotubes (CNT; modelled)	$10^{-5} - 10^{-3}$	$10^{-2} - 2 \times 10^{0}$
fullerenes (modelled)	$2 \times 10^{-5} - 10^{-4}$	$7 \times 10^{-4} - 2 \times 10^{-1}$
CeO_2 (modelled)	$7 \times 10^{-3} - 7 \times 10^{-1}$	–

3 Small-Scale Modelling

Crucial to most if not all modelling techniques is the idea that some form of conservation law applies. Often it is the mass of a substance that must conform to a balance equation. If a substance leaves one compartment due to transport, then the mass that is removed from that compartment must be added to that in the receiving compartment, unless some chemical process transforms the substance into a different substance. Population balance models, however, deal with the number of particles (the population) of similar size and composition instead of the mass. With this type of models the number of particles evolves according to a conservation law. Particles can aggregate into larger particles, thereby changing to a different size—the number of small particles is reduced and the number of large particles is increased.

Both the number of particles and the mass of the particles are useful measures, but they require different modelling approaches. As discussed in Sect. 6.3, which measure is the most appropriate for quantifying the exposure is a matter of debate (see also (Grieger et al. 2010)), but both model types can be used in combination with a hydrological model to predict the transport of nanoparticles.

The strong point of this type of modelling is that it can help us better understand the processes that are involved. Some of the modelling techniques are directly applicable in larger-scale models, but some like the particle tracking require too much resources and instead the insight they provide has to be translated for the larger scale.

3.1 Population Balance Models

One of the processes that determine the fate of the nanoparticles in the environment, is the clustering of these particles, either forming homogeneous clusters, so-called homoaggregation, or clusters with clay and organic particles in the micrometer range, so-called heteroaggregation (see Fig. 2). Many authors seek to describe these processes in terms of the number of particles and clusters of a particular size, e.g. Degueldre et al. (2009).

The "free" nanoparticles and the nanoparticles in clusters are divided into size classes and equations are developed to describe the evolution of the number of particles and clusters in each size class [see Appendix A for details; (Quik et al. 2014)].

The collision rates that determine the formation and destruction of clusters of particles can be related to the properties of the particles and the surrounding water (Praetorius et al. 2012; Arvidsson et al. 2011). The equations can be extended to include such processes as sedimentation to or resuspension from the water bottom, as these processes also influence the number of particles in surface water systems.

In principle one needs to distinguish size classes of all sizes, but in practice a reasonable limit is chosen, based on the idea that ever larger clusters are very rare. The introduction of SPM like clay or organic particles to describe heteroaggregation makes the population balance equation more complicated and some simplifications must be made. Quik et al. (2014) for instance analysed the equations for possible simplifications and on this basis applied the assumption that only clusters of nanoparticles exceeding a critical size and nanoparticles attached to SPM, are subject to sedimentation. This simplifies the mathematical model, as now essentially only two classes have to be distinguished. In a follow-up study, using a numerical model which could deal with the extra complexity, Quik et al. did distinguish several size classes (Quik et al. 2015).

Atmuri et al. (2013) combined the population balance approach with the DLVO theory (see Sect. 3.5), to predict different regimes of aggregation, at varying salt concentrations. Some tuning of the various model parameters was required and the aggregation rates they predicted were much faster than those observed.

3.2 *Mass Concentrations*

A drawback of population balance models is that they involve a large number of parameters (the collision rate coefficients for the various interacting classes), even if simplifying assumptions like one value for all collision rate coefficients k_{ij} can be made, but also require arbitrary choices, such as the number of size classes to distinguish. A further drawback is that the number concentration is much less intuitive than the mass concentration. In some areas of environmental science the number concentration is widely used, for instance in atmospheric pollution studies, but it is not often used in connection with the aquatic environment.

The classical representation of concentrations as mass concentrations can, however, be used for nanoparticles just as it is used for dissolved substances. For instance, Brunelli et al. (2013) used a first-order differential equation, $dC/dt = -kC$, to find the sedimentation rate coefficient k of titanium dioxide nanoparticles in the diverse natural and synthetic waters in their study. The coefficient thus obtained is specific for each experiment. The results can not readily be applied to other experiments or natural systems.

Markus et al. (2015) used a set of differential equations linking the mass concentrations of various fractions of nanoparticles to develop a more extensive model of the sedimentation and aggregation processes. They applied it to published laboratory experiments and derived a more or less universal set of process parameters, so that application to other systems is possible.

3.3 Dissolution

An important property of nanoparticles is the solubility of the material, as it determines the rate at which ions are released and therefore determines at least part of the toxicity (Beer et al. 2012). The solubility depends on the material the nanoparticles are made of, including the coating, and can be influenced by other chemical transformations, such as the formation of sulfides in the case of silver (Dale et al. 2013; Levard et al. 2012). Both silver and zinc oxide nanoparticles are known to dissolve to some degree. Other commonly used inorganic nanoparticles, such as cerium dioxide and titanium dioxide, are much less subjected to chemical transformations or dissolution.

Dissolution of silver nanoparticles turns out to be a complex process, which involves the oxidation of silver to silver ions and may lead to the formation of silver sulfide (Levard et al. 2012; Liu et al. 2011a). In general the rate at which nanoparticles dissolve depends on the size, as shown experimentally by David et al. (2012), as well as Zhang et al. (2011). The use of the Oswald-Freundlich equation, which relates the solubility to the curvature of the particles' surface, to explain the size-dependence was criticised by Kaptay (2012). The thermodynamic exposé he presents shows that the effect is caused by the specific surface area instead. Mihranyan and Strømme nevertheless invoked this same equation to study the solubility of nanoparticles with a rough surface, instead of the often assumed ideal spherical shape (Mihranyan and Strømme 2007). They found that the solubility is significantly enhanced with respect to the classical theory.

Zhang et al. (2010) studied the effect of size on the dissolution of zinc sulfide nanoparticles, but they also included the effects of pH. The net effect is rather complicated, as it is influenced by the chemistry of the nanoparticles and the coatings that are applied. What is clear, however, is that both particle size and pH strongly influence the dissolution. To explain the effects, Zhang et al. used a thermodynamic analysis.

David et al. (2012), Zhang et al. (2011), (2010) have presented data on the dissolution kinetics of zinc oxide, silver and zinc sulfide, respectively, that show a timescale of one to several hours before the equilibrium concentration is reached. This means that dissolution is a fast process, at least in the circumstances they studied.

In contrast to these findings, Lowry et al. (2012) mention that in their constructed wetland not all silver is transformed into silver sulfide, even after 18 months. Some 30 % is still bio-available. Thus the time scale for these processes can vary widely and there is little understanding what causes the difference.

3.4 Chemical and Physical Reactivity

As already mentioned in the introduction, to date few, if any, models are capable of handling the various environmental factors that influence the fate of nanoparticles (Dale et al. 2015a). This includes the pH and the ionic strength of the water. Work by Kaegi et al. (2011), Brunetti et al. (2015) and Dale et al. (2013) has shown the importance of sulfidation for silver nanoparticles. As silver sulfide is virtually insoluble in water, whereas silver is soluble, this has consequences for the fate of silver nanoparticles. Due to sulfidation a shell is formed around silver nanoparticles and the nanoparticles persist as particles, whereas zinc oxide nanoparticles may completely dissolve.

Other aspects of the interactions between nanoparticles and their surroundings, such as the photocatalytic properties, receive only little attention in the context of modelling. As an exception, Hotze et al. (2010) developed a framework for predicting the reactivity of aggregates of nanoparticles, for instance the generation of reactive oxygen species (ROS) by fullerenes and their derivatives. According to their theory aggregates of nanoparticles may be much more reactive than one would expect from measurements of the ROS generation by separate nanoparticles. As nanoparticles mostly exist as aggregates, this means that for the proper modelling of ROS generation aggregates will have to be treated separately.

3.5 DLVO Theory

The most commonly used theory to explain the aggregation behaviour of nanoparticles is the classical DLVO theory developed by Derjaguin and Landau and independently by Verwey and Overbeek, in the 1940s, with or without extensions specific to nanoparticles (Wikipedia 2015). Petosa et al. (2010) provide an overview of the different approaches to describe aggregation and deposition of nanoparticles, including the various phenomena that are not considered in the classical theory. In their publication deposition is to be understood as the deposition of nanoparticles on macroscopic surfaces or the surfaces of particles in the micro- and millimeter range.

In its simplest form the DLVO theory predicts the potential energy between a colloidal particle and a (macroscopic) surface or between two colloidal particles as the sum of electrostatic and van der Waals forces. The two parameters in this theory, the Debye-Hückel length and the Hamaker constant, both depend on the

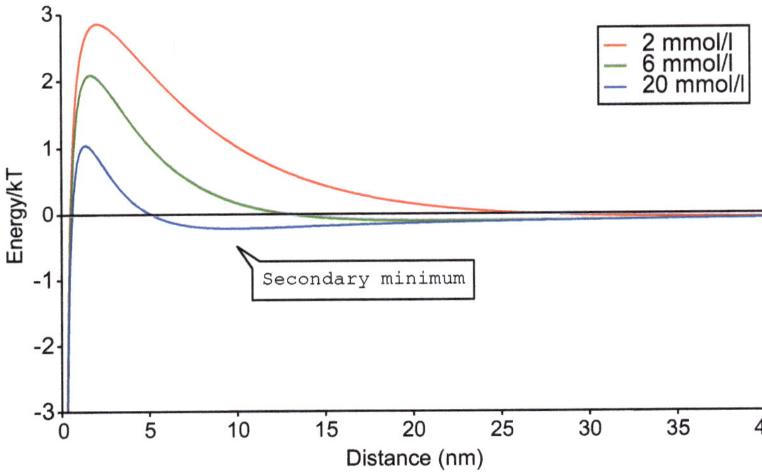

Fig. 3 Illustration of the interaction energy according to the DLVO theory, with different ionic strengths. The secondary minimum, visible in one curve, can cause the particles to form loose aggregates. The inverse Debye-Hückel length was set to $\sqrt{(I)}/0.3\,\text{nm}$, I being the ionic strength in mol/l.

ionic strength of the medium. The Hamaker constant also depends on the characteristics of the colloidal particles and the surfaces in question. The theory is used to examine if there is a minimum in the potential energy, which indicates whether the colloidal particles remain separated or instead aggregate in this minimum (see Fig. 3).

While DLVO theory generally includes these two forces to describe the stability of a suspension of nanoparticles, one frequently needs to deal with magnetic forces, e.g. when iron nanoparticles are involved, steric interactions as well as hydration forces. These effects lead to extra terms in the expression for the potential energy.

Magnetic forces in particular have been studied by Phenrat et al. (2007). Their experimental evidence confirms that the magnetic attractive forces of iron nanoparticles cause an enhanced aggregration compared to non-magnetic nanoparticles. This has consequences for the application of such particles in soil remediation projects. They recommend to use the aggregation size rather the size of individual particles to characterize the nanoparticles in analyses.

Li and Chen (2012) applied an extended form of the DLVO theory to their experiments with the aggregation of cerium dioxide nanoparticles in aqueous solutions with varying ionic strengths and concentrations of humic acids. They concluded that this extended theory fits the experimental results well. Of particular interest is that humic acids in some circumstances appear to hinder aggregation but in other circumstances promote aggregation. The latter occurs when a sufficiently high concentration of a divalent electrolyte is present.

Besides postulating the form that the potential energy expression should have, given the various forces acting on the nanoparticles and determining from that the behaviour of a suspension, one can also simulate the behaviour of collections of nanoparticles (see also Sect. 3.6). Two publications relating the DLVO theory to such simulations are: Macpherson et al. (Macpherson et al. 2012) who investigate the effect of the Hamaker constant and the particle concentration on the aggregates that are formed and Dahirel and Jardat (2010) whose review describes under which circumstances DLVO theory is not an adequate approach. The particle tracking techniques on which these articles rely may in principle allow us a better understanding of the aggregation behaviour, but they also show that the details of the process depend on many aspects of the medium and of the nanoparticles in question. At present particle tracking is not suitable for studying practical questions.

3.6 Particle Dynamics

Several publications have been dedicated to the structure of the aggregates that are formed during the aggregation process with the intention of better understanding the properties of these aggregates and how they are related to the properties of the nanoparticles. Two modelling approaches are described:

- Model the motion of individual nanoparticles as a consequence of Brownian motion and the forces between the nanoparticles, for which the DLVO theory with or without extensions is commonly used. An example of this approach is given by Peng et al. (2010) and to a lesser extent, as it focuses on particles of micrometer scale, by Satoh and Taneko (2009). In both cases simulation of the motion of individual particles and the subsequent formation of aggregates is used to determine how fast the aggregates are formed and what the size and other characteristics are.

  Tracking individual particles is a computationally intense method, since large numbers of particles are involved with interactions between pairs of particles or aggregates. Liu et al. (2011b) mention that for their simulations concerning a population of 10,000 particles, where new particles were inserted into simulation after an aggregate was formed to keep the population at the same number, the computer required 15 days to complete the task.

  An alternative mathematical approach, used by Lattuada et al. (2006), seeks to describe the dynamics not in terms of individual particles, but rather as size populations. While this approach is closely related to the population balance models (Sect. 3.1), they emphasize the process of aggregation as influenced by the forces between the particles, not the effects on the size distribution.
- Use the concept of fractals to characterize the geometry of the aggregates—a large fractal dimension means that the aggregates are compact, whereas a small fractal dimension indicates a loose structure. This has consequences for the reactivity of the aggregates, as the available surface area depends on the geometry. Zhou and Keller (2010) used this concept to determine the character of

aggregates that result in the so-called reaction-limited regime of aggregation (RLCA) and the diffusion-limited regime (DLCA), where they used both spherical and irregularly shaped nanoparticles.

Furthermore, the concept of a fractal dimension was used by Di et al. to describe the effect of aggregation on the dissolution of silver nanoparticles (He et al. 2013). This resulted in a model to predict the release of silver ions under different environmental circumstances, which is important for the prediction of the toxicity of this type of nanoparticles to organisms.

4 Intermediate Scale: Environmental and Technical Compartments

The second category of models holds the middle ground between small-scale modelling techniques, concerned with the details of the physico-chemical processes, and large-scale techniques where these processes are described mostly with semi-empirical relations. The approach of this category of models is to consider the processes to which the nanomaterials are subject, using the insights obtained from small-scale modelling, as described in the previous section, in conjunction with the detailed hydrology or hydrodynamics of the system in question.

4.1 Groundwater and Soil

Various authors, for instance Müller et al. (2013) and Gottschalk et al. (2013), have concluded that most nanoparticles end up in landfills, because the ash left after sewage has been incinerated will be stored there, or in soil, when sewage sludge is used as fertilizer. Thus, understanding the fate of nanoparticles in these environmental compartments is of paramount importance. Quite a few publications have been devoted to this problem, but the majority focuses on the behaviour in idealized soils, for instance glass beads or clean sand (Ben-Moshe et al. 2010; Li et al. 2011). While such artificial soils are much easier to define and characterise, the question is whether they can indeed be representative for what happens in real soils. Also the preparation of the columns in which the experiments are performed may influence the results: columns made out of glass beads will not have the sort of inhomogeneities that real soils have, such as cracks, inclusions of other material and so on (Fang et al. 2009).

When it comes to modelling the transport of nanoparticles in soils, be they real or idealized, most authors use classical filtration theory to explain the form of the breakthrough curves and retention profiles—properties like the time the contaminants appear in the water that has passed through the column and the total amount of contaminants that actually leave the column versus the amount that was injected (Liang et al. 2013). The classical filtration theory is summarised in Appendix A.

The retention of nanoparticles, that is, either adsorption to the soil particles or blocking because the nanoparticles are too large for the canals between the soil particles, is most often described via an exchange between a mobile fraction and an adsorbed fraction. It is then assumed that the exchange is governed by first-order processes (adsorption or, as it is often also called, deposition to the soil particles and release into the water phase), whose rate coefficients have to be determined empirically (Li et al. 2011).

Little is known about the adsorption and desorption rate coefficients, but Tufenkji and Elimelech (2004) have developed a semi-empirical formula for the collision rates of colloidal and nano-sized particles with the soil as a consequence of various transport mechanisms. Unfortunately, to calculate the net attachment rate, one also needs to know the efficiency by which such collisions lead to a permanent attachment. Badawy et al. (2013) actually used the theoretical relations presented by Tufenkji and Elimelech to estimate the adsorption efficiency (see Appendix A).

Goldberg et al. (2014) published a critical overview of the various model formulations that have been used to explain the observed retention profiles and breakthrough curves. Their conclusion was that complicated models do not necessarily explain the laboratory experiments better than simpler models. A crucial point of concern is that the data obtained in the experiments must enable the determination of the various parameters in the model. Alternatively, the model formulation should allow experimental determination.

Besides column experiments several authors present studies of the transport in two or three dimensions. Bai and Li (2012) for instance studied the distribution of fullerenes (nC_{60}) using a groundwater model extended with the relevant attachment processes (MT3DMS) and Cullen et al. (2010) modelled the mobility of carbon-based nanoparticles (fullerenes and various types of carbon nanotubes) in a field with heterogeneous permeability to mimic the environmental conditions. The basic description of the processes involved is in both cases very similar to what is described above.

Areepitak and Ren (2011) and Boncagni et al. (2009) also used colloid filtration theory to describe the exchange of nanoparticles between a stream and its stream-bed. In this situation an additional complication exists, however, namely that water is exchanged between the stream and the bed under wave action. Just as in the column experiments discussed above the adsorption efficiency is an important parameter.

The NanoRem project takes a markedly different approach (NanoRem 2015). The purpose of this project is to further develop methods for soil remediation using nanoparticles like nano zero-valent iron. Besides experiments to elucidate the behaviour of nanoparticles in the soil and the ecotoxicological aspects the project involves modelling the behaviour of nanoparticles at pore level. The advantage of this scale is that fundamental physical laws can be applied to describe the behaviour of the particles. The phenomena that are observed on this microscopic scale are then to be translated to a macroscopic scale, so that they can be applied in a groundwater like MODFLOW (USGS 2015). This is work in progress.

4.2 Surface Waters

The current literature contains a small number of publications that focus on the modelling of the transport of nanoparticles in freshwater bodies. No publications were found that consider marine or estuarine water systems.

Praetorius et al. (2012) developed a model for titanium dioxide in the river Rhine, based on earlier work by Blaser et al. (2008). The hydrodynamics of the Rhine was modelled schematically, using a series of boxes, in which river flow, bed load transport and interaction with the sediment are distinguished. The processes nanoparticles are subject to, sedimentation, homoaggregation and hetero-aggregation, were modelled using a population balance approach (see Sect. 3.1 for some details and Fig. 2). Crucial in this model is the efficiency of the heteroaggregation process. To gain insight in its effect on the distribution of the titanium dioxide nanoparticles, they used a range of values, 0.001–1, for the aggregation efficiency in their model simulations. Other influences on the aggregation process, were modelled explicitly: Brownian motion, fluid motion and differential settling (Elimelich et al. 1998).

The same modelling technique was applied to the lower Rhône river in France by Sani-Kast et al. (2015). Their purpose was to overcome some limitations of the models to date, namely the lack of spatial and temporal variability and to analyse the effect of widely varying environmental conditions, that is, variations in the water chemistry. To this end the model was run with a wide range of conditions and the results were analysed using cluster analysis. Heteroaggregation was found to be a significant factor: if the circumstances favour the attachment of nanoparticles to SPM, then other factors like the water chemistry had little influence on the transport and fate.

A different approach was taken by Dale et al. (2013) in their study of the fate of silver nanoparticles in freshwater sediments. The objective was to develop a model that can predict the distribution of these nanoparticles in the sediment and the speciation of the silver. Chemical transformation of silver into silver sulfide is a relevant environmental process, because it influences the toxicity of the silver: silver sulfide is almost insoluble and therefore not biologically available, whereas metallic silver and especially silver ions are. The model they developed predicts that the resulting coating of the nanoparticles with silver sulfide reduces the release of silver ions. In a later publication (Dale et al. 2015b) they elaborated on this work for other types of nanoparticles, such as the chemical transformations that zinc oxide nanoparticles are subjected to.

Quik et al. (2015) used a population balance model with five size classes of both engineered nanoparticles and natural suspended solids to describe the transport of nanoparticles in rivers, using a Dutch river for a case study (see also Sect. 3.1). Their main conclusion is that both spatial heterogeneity and particle size distribution should be explicitly modelled. They found that ignoring this factor could lead to an underestimate of the concentrations in the sediment by a factor 20.

In a follow-up study from (Markus et al. 2015, 2016) describe the application of their model for the aggregation and sedimentation processes to the transport and fate of metallic nanoparticles (zinc oxide, titanium dioxide and silver) in the river Rhine (Markus et al. 2015). The basis for this work was a hydraulic model of the Rhine from Basel in Switzerland up to the central parts of the Netherlands. They studied several scenarios regarding the release of nanoparticles through wastewater: one assuming only release via the wastewater treatment plants and others assuming an additional diffuse emission due to leaching of nanoparticles from land-applied sewage sludge. The model calculations showed that the contribution of zinc oxide nanoparticles in terms of mass is likely in the order of 5–10 % of the observed zinc concentrations.

4.3 Wastewater Treatment Plants and Other Technical Compartments

The so-called technical compartments, wastewater treatment plants (WWTP), waste incineration plants (WIP) and others, are a subject of study for several reasons. First of all, they form a buffer between the waste stream from households and industry on the one hand and the environment on the other. The most important aspects are then: can they filter out the engineered nanoparticles and to what extent does the passage alter the character of these materials? Secondly, one can actively use nanoparticles to clean the wastewater, for instance using the photocatalytic properties of titanium dioxide nanoparticles to reduce organic micropollutants (Mohapatra et al. 2014b).

Benn and Westerhoff (2008) investigated the release of silver from commercial textiles and the fate of this silver in a WWTP. They used a straightforward mass-balance model with non-linear sorption to predict the distribution of the silver over effluent and sewage. The conclusion was that much of the silver will be retained in the sewage sludge, which as a consequence may no longer be suitable as a fertilizer. Note, however, the critique by Praetorius et al. (2014) as this model assumed a thermodynamic equilibrium (see Sect. 2).

A comparable approach was taken by Yang et al. (2015) to analyse the experimental results they obtained with a sequential batch reactor (SBR) to simultaneously remove nanosilver and fullerenes from wastewaster. In the experiments 95 % of the fullerenes and 90 % of the silver was removed. The authors claimed that the distribution coefficients they found can be used to predict the removal of nanoparticles in such installations.

Mahmoodi et al. (2008) used titanium dioxide nanoparticles, fixed to polymer membranes, to degrade an organic pollutant of agricultural origin, fenitrothion, via the photocatalytic action of these nanoparticles. The degradation process was modelled using first-order decay. Their full model was one-dimensional, in order to describe the complete experimental set-up, but they made no effort to model the photocatalysis explicitly.

The role of waste incineration in the fate of nanoparticles has been discussed above (see Sect. 5.2). Roes et al. (2012) indicates that not all nanoparticles will actually be caught in the residue but will instead escape via the off-gas. This holds especially for particles smaller than 100 nm. The authors use a conceptual model of WIPs and results from previous studies, combined with technical details of the various types of WIPs in use to arrive at their conclusions.

4.4 Probabilistic Methods

In general the fate of nanoparticles in natural waters or in technical installations, like wastewater treatment plants is poorly understood, as a consequence of the difficulty of measuring the concentrations and the wide variety of circumstances that influence the processes. In order to deal with these uncertainties probabilistic methods are often used, where inputs into the system and process coefficients are varied, to gain insight in the range of emissions and concentrations.

In the literature one can find at least two categories of probabilistic modelling approaches that are of interest here: Monte Carlo simulations (MC) and "direct" probabilistic modelling. MC simulations are often used to quantify the distribution of the output parameters the authors are interested in. For instance Barton et al. (2015) used the technique to estimate the removal of three types of metal and metal oxide nanoparticles in wastewater treatment. The reason for using an MC technique is that the parameters involved are not known with any precision. They applied a large number of random values for these parameters and then determined a probability distribution of the emission of nanoparticles (see also Sect. 5.2).

Hendren et al. (2013b) also used this technique for modelling the fate of silver nanoparticles in wastewater treatment. One characteristic of this approach is that one needs to assume a reasonable distribution for the various parameters, so that the output is more or less realistic. Choosing such a distribution may not be trivial (see also Section and the analysis by Ferson et al. (2001).

Probabilistic modelling can take a different form as well, as illustrated by Jacobs et al. (2015). They used the results of a previous deterministic study regarding the use of silica nanoparticles in food to quantify the various sources of uncertainty. The most important were the uncertainty in the dose causing effects (the benchmark dose), variation among individuals and the extrapolation from subchronic to chronic effects. All parameters do not have an equal contribution to the total uncertainty and by focussing on the most important parameters efforts to reduce the uncertainty become more efficient.

5 Large-Scale Modelling

As stated in the introduction, several techniques are used for answering or analysing regional or even global issues. They have in common that the transport of contaminants between the various parts of the system under study is described via general concepts such as transfer coefficients instead of via hydrodynamic or hydrological models (Mackay et al. 2001). Geographical details are often "lumped," which leads to the representation of the system via a number of interacting "boxes," all with their own characteristics. The exception to this is the use of GIS as a modelling tool.

5.1 *Multimedia Models*

The purpose of multimedia models is to determine the distribution of a contaminant, in this case of nanoparticles, over various environmental compartments but often also over so-called technical compartments. The latter type is used to be able to deal with relevant industrial processes, such as the production of nanoparticles or the effects of wastewater treatment on the release into the environment. In the context of this review environmental compartments include: surface water (mostly inland waters), effluent, sludge, air and soil (see Fig. 4).

The interaction of one compartment with another is often very complex. For instance incineration of sewage sludge may bring nanoparticles into the atmosphere (Buha et al. 2014), so that they are spread via the wind but later they are deposited on the soil due to rainfall or sedimentation of the particles. Taking all details into account is an almost impossible task and therefore multimedia models often

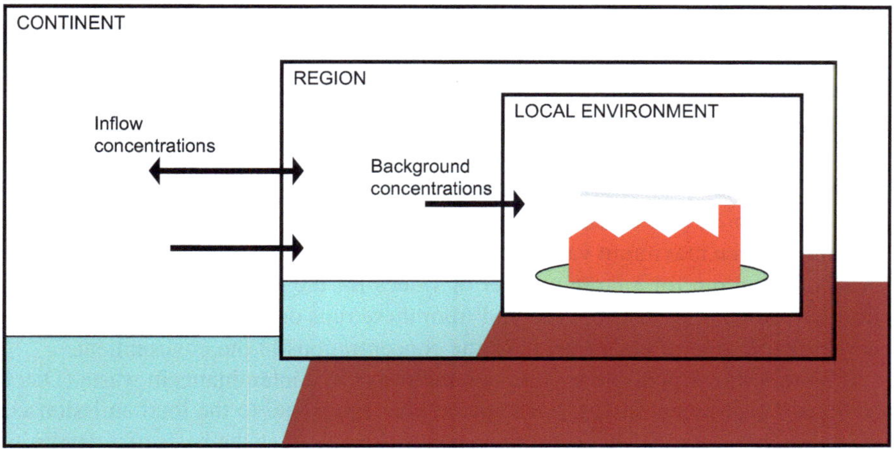

Fig. 4 Example of the set-up used in multimedia modelling. The area of interest is conceptually divided into "boxes," called "unit worlds" in Mackay et al. (2001), between which matter is exchanged. Adapted from (Mackay et al. 2001)

aggregate the details into a small set of exchange processes that work on "generic" compartments. Instead of describing every single river or agricultural area, a generic river and a typical piece of arable land are considered. A mass balance calculation is then used to calculate the final (average) concentrations that occur in each compartment as a function of the inputs into the whole system.

Within the literature specifically concerned with nanoparticles two articles in particular apply this type of modelling: Meesters et al. (2014) and Liu and Cohen (2014). The models differ in the emphasis on particular processes: Meesters et al. use a mechanistic approach, so that aggregation in surface water and deposition through rain on soil are explicitly described, while Liu and Cohen rely exclusively on exchange coefficients.

Some authors argue that the processes that need to be taken into account can be described via simple first-order expressions (Quik et al. 2011). The main argument for this approach is that to date there is too little quantitative information to justify a more complicated description.

Multimedia models are particularly suited for screening purposes: to identify where contaminants will end up and to estimate at least the order of magnitude of the concentration in the entire environment. If one seeks a more or less global approach, then multimedia models are certainly a good candidate (Westerhoff and Nowack 2013).

5.2 *Material Flow Analysis and Exposure Modelling*

The goals of material flow analysis (MFA) are less all-embracing than those of multimedia models in the sense that the focus is on tracing mass flows for instance from wastewater produced by households to receiving surface waters, instead of considering all environmental compartments. But in common with multimedia models MFA uses transfer coefficients to describe the flow of contaminants from one compartment to the next. More emphasis is put on dealing with the uncertainties in the transfer coefficients and emission estimates by using probabilistic techniques (Sun et al. 2014; Gottschalk et al. 2010a, b).

Basically, for each transfer coefficient a likely probability distribution is estimated, which can be as straightforward as a uniform distribution, defined by a minimum and a maximum value or which can have a more complicated form. The model is then run for a large number of values for the transfer coefficients drawn from the probability distributions and from these runs one can determine the shape of the distribution of emissions to or the concentrations in the environment.

Figure 5 shows possible sources and pathways of contaminants in waste (Dutch National Government 2015). Each source may contribute to the total emission via the given pathways. Such a schematisation may be used as the basis of a deterministic calculation, if sufficient knowledge is available for estimates, but also a probabilistic calculation (see also the discussion in Sect. 6.1). In essence the various sources of nanoparticles or other contaminants are traced via recognised pathways

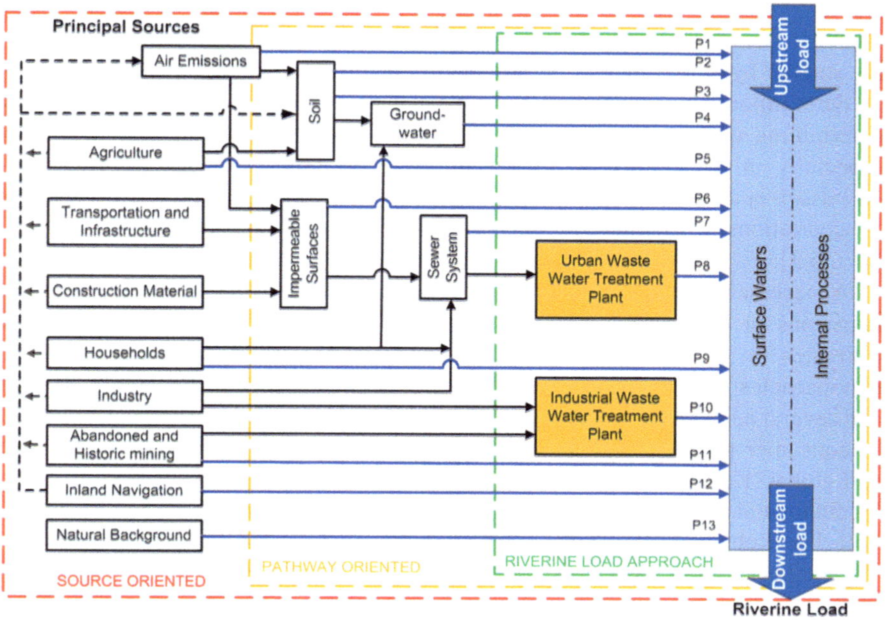

Fig. 5 Sketch of the contributions and pathways of wastewater containing nanoparticles (Dutch National Government 2015)

to the receiving aquatic environment, making the estimation of the total emission possible.

Müller et al. (2013) used such a probabilistic technique to gain insight in the fate of various types of engineered nanoparticles during waste handling processes such as the incineration of sewage sludge. Buha et al. (2014) compared their model results based on probabilistic MFA with their measurements of particle sizes in fly ash. Their study demonstrated that nanoparticles may survive the incineration process and even that new nano-sized particles may be formed.

Koehler et al. (2011) considered the incineration process in greater detail, showing that these models are not limited to global set-ups but can be used for detailed analyses. That does not mean, however, that such models are applicable everywhere: the transfer coefficients the model relies on have to be measured or estimated in some way and often the lack of detailed information will hinder such estimation. If the transfer of material from one compartment to another cannot be described as a linear process, the modelling needs to be adjusted to accommodate an alternative mathematical description or a different approach needs to be taken.

Müller et al. (2013) included both waste incineration plants and landfills in their study. While some aspects are specific to the situation in Switzerland, notably the fact that almost all sewage sludge is incinerated, their analysis predicted that most engineered nanomaterial will end up in the slag resulting from the incineration process and therefore eventually in landfills. Other mass flows were smaller by at least an order of magnitude. Of the nanoparticles they examined, titanium dioxide is the most important one—the mass flows of zinc oxide, silver and carbon nanotubes are smaller by at least one or two orders of magnitude.

Furthermore, they concluded that most data in the literature relate to *mass* as the determining quantity, whereas sometimes the *number of particles* is preferred, especially in the context of ecotoxicology. This point was also addressed by Arvidsson et al. (2012), who used a variant of MFA to directly predict the number concentrations of titanium dioxide nanoparticles [see also (Arvidsson 2012) and Sect. 6.3].

A necessary input for material flow analysis to provide an estimate of the emissions of nanoparticles to any environmental compartment is that we have estimates or, preferably, tangible data about the number or concentration of nanoparticles or any other contaminant that is being put into the waste. To this end several authors have attempted to make an inventory of the use of nanoparticles in consumer products and industrial processes at various scales: local, regional and global (Keller and Lazareva 2014; Robichaud et al. 2009; Lorenz et al. 2011; Piccinno et al. 2012). Some of these efforts consider only a single type of nanoparticle, like Robichaud et al. (2009), who focussed on titanium dioxide nanoparticles, while others consider a very broad spectrum, like Lorenz et al. (2011), who identified the possible exposure of consumers in Germany to engineered nanoparticles in general.

The problem is that no reliable sources of information on the actual use of nanoparticles in consumer products exist. Even a much quoted database as the one from the Woodrow Wilson Institute is inaccurate and incomplete (Berube et al. 2010). This is partly due to the listed products having a short life time—quite often one cannot find a particular product anymore—and partly due to the manufacturers not providing information. Boxall et al. (2007) provided some of the more comprehensive estimates published to date, but this remains a significant problem.

Publications such as by Boxall et al. (2007) but also by Lorenz et al. (2011) enabled Markus et al. to predict the relative contribution of engineered nanoparticles to the total metal load of the Rhine and Meuse rivers (Markus et al. 2013). Their conclusion was that based on the available information engineered nanoparticles released within the Netherlands might contribute 5–10 % to the total load of these metals, in terms of mass.

5.3 GIS Approaches

Geographical information systems (GIS) are frequently used for analysing environmental problems, for instance Hüffmeyer et al. (2009) and Comber et al. (2013), but they do not seem very popular in relation to nanoparticles. In the European project NanoFATE water bodies have been classified by their hydrochemical characteristics (Hammes et al. 2013). First the water bodies were examined for similarities between pH, alkalinity, electric conductivity and other parameters, using principle components analysis. The analysis led to the definition of six classes of water bodies, which especially differ in ionic strength and organic content. All water bodies were assigned to one of these classes. The classification was then used to predict the fate of nanoparticles, that is, predict the tendency of colloidal and nano-scale particles to aggregate. Generally speaking, a high ionic strength means the particles are likely to aggregate and a low ionic strength means that the suspension is more stable. The authors used a georeferenced database containing chemical information about 800 European rivers for their analysis.

A different approach was taken by Dumont et al. (2015), who used a combination of a GIS and a hydrological model, called GWAVA, with a water quality component. The latter was used to calculate the concentrations of silver and zinc oxide nanoparticles from the emission of nanoparticles as a function of the population density. Akin to this is the use of the GREAT-ER model by Kehrein et al. (2015), although they focussed on micropollutants, rather than nanoparticles.

The combination of GIS systems with associated databases regarding population, land use and such, is a powerful one, especially if the relevant hydrological information is available, for instance via the development of models for nutrients or other contaminants.

6 Conclusions and Summary

The collection of articles on the modelling of nanoparticles in the environment, as presented here, shows that many different approaches are used to answer the question of how to predict the release, transport and fate of engineered nanoparticles in the environment (Table 3). An important motivation for these modelling efforts is that we lack actual empirical data on the occurrence and fate of these contaminants, except under controlled circumstances.

Each of the modelling approaches described in the literature has its pros and cons. For instance, if the problem is to determine the distribution of nanoparticles produced by traffic over air and soil, then—due to the spatial scale—it is not feasible, given current computational resources, to use a model that tracks the position of individual nanoparticles as a consequence of air flow and Brownian motion. That would require tracing the position of billions of particles over a considerable period. Instead a multimedia model would be a more suitable choice. However, such a model is not suitable for answering questions about the size distribution of aggregates of nanoparticles. That is a question that might arise when considering the potential consequences of nanoparticles for bottom-dwelling organisms (Schaumann et al. 2015). Large aggregates will quickly sink to the bottom of a lake whereas small aggregates and individual nanoparticles are likely to stay in suspension for a long time.

Table 3 Overview of the modelling techniques and their predictive capabilities

Modelling technique	Predictions, remarks
Small-scale	Describe processes (fate) in models for transport
Population balance	Number concentration and size distribution. Chemical transformation processes and dissolution require special attention
Mass concentrations	Mass concentration, processes like dissolution easily modelled (mass conservation)
DLVO theory	Useful for understanding aggregation. Complications arise for particular types of nanoparticles, reducing practical usefulness
Particle dynamics	Useful for understanding aggregation Computationally intensive
Intermediate-scale	Models for transport based on hydrodynamics or hydrology, mostly limited to one medium, but especially suitable for specific geographical areas Chemical and physical processes based on small-scale techniques
Groundwater models	Spreading of nanoparticles in groundwater, released from sewage sludge or used for remediation
Surface water models	Spreading of nanoparticles in surface water Consequences of release into the environment
Technical compartments	Estimation of distribution over sludge and effluent Fate in waste incineration plants—distribution over ash and out-gas (atmospheric pollution)
Large-scale	Models for transport based on "lumped" transfer processes Chemical and physical processes based on
Multimedia models	Designed for predicting the fate of contaminants in several media Especially useful for screening/regulatory purposes Limited attention for transport details Chemical and physical processes based on small-scale techniques
Material flow analysis	Easily combined with probabilistic techniques, providing insight in consequences of uncertainty
GIS modelling	Process large amounts of data, mapping the results No or very limited interaction between geographical elements

6.1 Data on Use, Production and Emission

We do not have reliable data on the production volumes, the application in consumer products or the subsequent release into the various environmental compartments. All we have are estimates of production and use and, based on experiments on laboratory and pilot scale, insights in transport of nanoparticles through wastewater treatment plants and waste incineration plants.

The method of material flow analysis is often used to examine the mass flows through a natural or technical compartment, like waste incineration plants or landfills (Sect. 5.2). This presumes at least some knowledge of the amounts of nanoparticles involved, however. Multimedia models encompass more than one such compartment and again rely on some knowledge of emissions (Sect. 5.1).

As mentioned in Sect. 5.2, material flow analysis is often combined with probabilistic techniques. To illustrate that, a simplified schematisation, based on Fig. 5 has been used, together with some fairly arbitrary numbers:

- Only sources and pathways via the sewer system or the surface water are considered.
- All sources are assumed to be known exactly, but the percentage of wastewater that is being treated, f_{treated} is assumed to be between 70 and 90 %.
- Similarly, the wastewater treatment plant retains a large fraction of the nanoparticles in the sewage sludge, thereby reducing the emission to the surface water. This fraction is assumed to be between 85 and 95 % (f_{treated})
- A fraction of the sewage sludge, between 15 and 25 % is assumed to be used as fertilizer (f_{fert}) and 15–25 % of the nanoparticles contained in the sludge is assumed to eventually leach into the surface water (f_{leach}).

The total fraction f_{release} of the nanoparticles in the original wastewater that enters the aquatic environment is:

$$f_{\text{release}} = f_{\text{treated}} \times (1 - f_{\text{retained}}) + (1 - f_{\text{treated}}) + f_{\text{retained}} \times f_{\text{fert}} \times f_{\text{leach}} \qquad (1)$$

If for each of the fraction a uniform distribution is used, this leads to a probability density like the one given in Fig. 6. The assumption of a uniform distribution may be replaced of course by any other suitable distribution, but the very use of (continuous) distributions gives a distinctly shaped graph that suggests more insight than is actually warranted.

While the uniform distribution is often regarded as the simplest possible, Ferson et al. (2001) argue that even with this distribution implicit assumptions are made that have to be justified with actual data or theoretical analyses. The alternative they propose is to analyse the *bounds* on the various output parameters that result from the bounds of the input parameters. This way implicit assumptions about statistical distributions can be avoided, while retaining all the information that is actually available. In the above example the result would be a range of 0.17–0.46 instead of the skewed graph.

What these methods have in common is that they take a "generic" approach: not a particular production line or a specific incineration plant, but an average production system or an average incineration plant. This leads to generally applicable results and insights, but for a given natural system or installation, things may turn out quite differently. Note that this "generic" approach is of a practical nature, rather than a fundamental limitation.

Modelling efforts using GIS systems have the advantage that they can provide regional and even local details about population density or land use to estimate the emission to a river system in detail, for instance. Such estimates can be used to feed a hydrological model for the river system at hand.

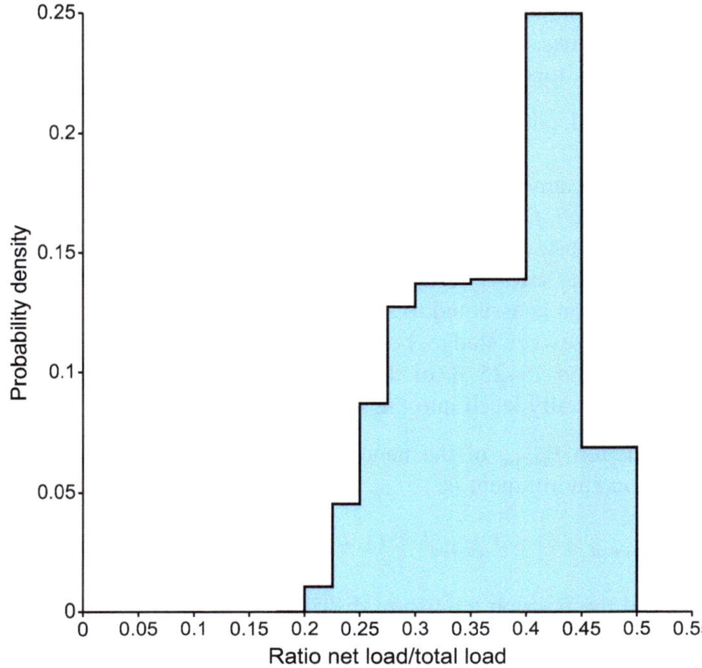

Fig. 6 Probability distribution of the fraction of the total emitted to the aquatic environment, calculated based on Eq. (1)

6.2 Modelling Processes in the Environment

Once the nanoparticles—of widely different shapes, sizes and materials—enter the environment, they are subject to a whole range of processes. Some are important only for certain types of nanoparticles, such as dissolution and chemical transformations, others are quite general—aggregation to SPM for instance. One can study these processes in laboratory settings and derive a mathematical model, but actual field data are required to verify that such a model is indeed adequate. A related problem is that coatings that have been applied for the proper functionalisation, and environmental factors like pH, the presence of oxygen, sulfide and natural organic matter as well as ionic strength all have influence on the fate of nanoparticles, but quantifying these influences turns out to be beyond our present capabilities.

That does not mean that nothing is possible. On the contrary, the various publications referred to in this paper show that we know a lot about the fate of nanoparticles. The models that are currently available, however, do not handle several important details, as indicated above. Approaches based on differential equations for the concentration and approaches based on particle population dynamics are both suitable as an add-on to surface and groundwater models.

Choosing the right parameters to describe the processes may be a challenge, given the variety of environmental factors, but general techniques like probabilistic modelling and sensitivity analysis will help.

6.3 *Open Questions and Guidelines*

Notwithstanding the variety of modelling studies and their findings as well as the available experimental studies, several important questions remain to be answered.

While some researchers prefer to use number concentrations, others work with mass concentrations. There is currently no clarity about which is to be preferred. It is not even clear if another measure should be used, such as the total surface area (Grieger et al. 2010):

- For silver nanoparticles consensus is that silver ions are the cause of the toxicity. Therefore a suitable measure would be the mass concentration.
- On the other hand, if the toxic effect of nanoparticles is due to disruption of the cell wall, as is reportedly the case of such materials as carbon nanotubes, then the number concentration seems more appropriate. This is also the measure that is used most often in the context of atmospheric pollution.
- But if the toxic effect is caused by the formation of reactive oxygen species at the surface of the particles, the total surface area might serve a better role.
- Some measurement techniques, such as single-particle ICP-MS or electron microscopy, are more naturally related to number concentrations than to mass concentrations. It might therefore be advantageous to further explore modelling techniques that connect to such measurements.

Currently there is a bias towards mass as the relevant measure, as noted by Müller et al. (2013). This is true for modelling as well as for measurements and published data. A conversion between the three measures is possible if the size distribution is known. To illustrate this, consider a set of nanoparticles ranging from almost zero to 200 nm. The number distribution is assumed to be uniform over this range. Then:

- Half of the particles will be below 100 nm in size.
- The mass of these smaller particles represent is 1/64 of the total mass.
- The surface area of these smaller particles is 1/24 of the total surface area.

These fractions will differ for different size distributions, but the example gives some insight in the order of magnitude small particles will contribute to the various measures.

The fact that we have too little concrete data and therefore rely on a large number of assumptions, albeit reasonable ones, leads to the realisation that the consequent uncertainty will influence the predictions of our models and that we need to explicitly deal with it.

Fortunately the variety of model approaches makes it possible to study the subject from different perspectives, thus providing a multitude of insights. In a nutshell:

- For a "quick" scan of possible issues, use methods like multimedia models or material flow analysis, as in general they do not require detailed data about a particular area. They do require proper insight in the phenomena at play, such as how much material is typically exchanged between compartments.
- For a more specific study of the transport and fate, use modelling techniques based on differential equations for the mass concentration or population balances for the number concentration. The model set-up includes a schematic representation of the area under study as well as a comprehensive mathematical description of the processes that play a role.
- The choice between a representation via "particle populations" or via "mass concentrations" is, perhaps, a matter of convenience. Some processes are easier to represent mathematically via a population balance, such as aggregation, others via mass concentrations, e.g. dissolution. Guaranteeing conservation of mass is definitely easier via mass concentrations.
- Consider the use of GIS systems to provide the regional information needed. The underlying databases will need to provide the correct level of detail, of course, but the data are already organised with respect to the geography and that is what is needed for a schematisation.
- Besides the choice of modelling approach, three additional aspects of nanoparticles in the environment ought to be taken into consideration:
  - All data we have available on the release of nanoparticles into the environment and their behaviour exhibit substantial uncertainties, as discussed by Hendren et al. (2013a). These have to be accounted for somehow, for instance via probabilistic modelling techniques.
  - As argued by Wagner et al. (2014), it is far from trivial to distinguish between natural and engineered nanoparticles. Still this needs to be done, even if mathematical models are well suited to make the distinction in principle—it is in the input data that this difference must be made.
  - The ultimate goal of these models must be to answer the question whether the spreading of (engineered) nanoparticles into the environment is harmful or not. That means that we need to know more about their ecotoxicology and how to relate that to the concentrations that we find via modelling. This also means that we need to better understand the possible (chemical) transformations as well as the reactivity, e.g. the production of ROS under influence of UV irradiation.

The most important consideration is that the model approach matches the questions that need to be answered. As there is a multitude of techniques to chose from, it should well be possible to find one that indeed matches. It should also be clear that besides the obvious questions additional aspects need to be addressed: uncertainties in the releases to be used, uncertainties in the processes (both the process parameters and which processes actually play a role) and what quantity (number or mass concentration) is suitable for expressing the amount and the ecotoxicological effect of nanoparticles in the aquatic environment.

Acknowledgements This work is supported by NanoNextNL, a micro and nanotechnology programme of the Dutch Government with 130 partners.

A Mathematical Details

A.1 Population Balance Theory

The "free" nanoparticles and the nanoparticles in clusters are divided into size classes and equations are developed to describe the evolution of the number of particles and clusters in each size class (Quik et al. 2014):

$$\frac{dN_j}{dt} = \frac{1}{2}\sum_{i=1}^{i=j-1} k_{i,j-i}N_iN_j - N_j\sum_{i=1}^{i=\infty} k_{i,j}N_i \tag{2}$$

where:

$k_{i,j}$ rate coefficient for the (successful) collision of particles in size classes i and j

N_i concentration of particles in the size class i

The first term in this equation represents the formation of larger clusters from individual particles or smaller clusters. The second term represents the reduction in number of the particles and clusters due to the formation of these larger clusters. No provision is made here for the disintegration of these clusters.

A.2 DLVO Theory

In its simplest form the DLVO theory predicts the potential energy between a colloidal particle and a (macroscopic) surface or between two colloidal particles as the sum of electrostatic and van der Waals forces. If furthermore the particles are assumed to be identical and therefore have the same surface potential and radius, then the interaction energy can be expressed as (Wikipedia 2015; Macpherson et al. 2012):

$$W(h) = W_{vdW}(h) + W_{dl}(h) = -\frac{AR^*}{6\pi h} + 2\epsilon\epsilon_0 R\psi_0^2 e^{-\kappa h} \tag{3}$$

$$R^* = \frac{R_1 R_2}{R_1 + R_2} = \frac{1}{2}R \tag{4}$$

where:

A	the Hamaker constant
h	distance between the particles' surfaces
R	radius of the particles
ε_0	the electric permittivity of vacuum
ε	the dielectric constant of water
κ	the inverse Debye-Hückel length
ψ_0	the surface potential of the particles

In this equation the first term is the contribution of the van der Waals forces and the second term is the contribution of the electrostatic forces, as modelled via the double-layer theory (Macpherson et al. 2012). The Debye-Hückel length and the Hamaker constant both depend on the ionic strength of the medium. The Hamaker constant also depends on the characteristics of the colloidal particles and the surfaces in question. The theory is used to examine if there is a minimum in the potential energy, which indicates whether the colloidal particles remain separated or instead aggregate in this minimum (see Fig. 3).

A.3 Transport and Adsorption in Groundwater

The equations that link the concentration of nanoparticles in the porewater (C) to the concentration of nanoparticles retained in the soil (S) are:

$$\frac{\partial C}{\partial t} + \frac{\rho_b}{n}\frac{\partial S}{\partial t} - \frac{\partial}{\partial z}D\frac{\partial C}{\partial z} + v\frac{\partial C}{\partial z} = 0 \tag{5}$$

and:

$$\frac{\rho_b}{n}\frac{\partial S}{\partial t} = k_{att}C - k_{det}S \tag{6}$$

where:

ρ_b	soil bulk density
n	porosity
v	velocity of the porewater
C	concentration of nanoparticles in the porewater
S	concentration of adsorbed nanoparticles

D diffusion coefficient
k_{att} adsorption (attachment) rate coefficient
k_{det} desorption (detachment) rate coefficient

This model formulation allows an arbitrarily high concentration of adsorbed nanoparticles, whereas in reality the adsorption capacity is finite. To accommodate a limited adsorption capacity, a blocking function may be introduced which effectively reduces the rate of adsorption as a function of the concentration of adsorbed nanoparticles (Liang et al. 2013; Kasel et al. 2013). Experience with such experiments has shown that the adsorption often depends on the distance from the entrance, leading to expressions like:

$$\psi = \left(1 - \frac{S}{S_{max}}\right)\left(\frac{d_{50} + z}{d_{50}}\right)^{-\beta} \tag{7}$$

where:
S_{max} capacity (maximum concentration) for the adsorption (deposition)
d_{50} size of the soil particles
z distance to the entrance
β shape parameter

and k_{att} in Eq. (6) is replaced by $k_{att}\psi$.

Tufenkji and Elimelech developed the following semi-empirical formula for the collision rates of colloidal and nano-sized particles with the soil as a consequence of various transport mchanisms (Tufenkji and Elimelech 2004):

$$k_{att} \sim 10^{-3.25} {N_{LO}}^{0.51} {N_{E1}}^{-0.27} {N_{DL}}^{1.06} \tag{8}$$

where:
N_{LO} the London number, relating the Hamaker constant, the viscosity of the fluid, the flow velocity and the particle diameter
N_{E1} the first electrokinetic parameter, which depends on the surface charge of the particles
N_{DL} the ratio of the particle diameter and the Debye-Hückel length

References

Areepitak T, Ren J (2011) Environ Sci Tech 45:5614. doi:10.1021/es200586v
Arvidsson R (2012) Contributions to emission, exposure and risk assessment of nanomaterials. Ph.D. thesis, Chalmers University of Techonology, Gothenburg, Sweden
Arvidsson R, Molander S, Sandén BA, Hassellöv M (2011) Hum Ecol Risk Assess 17:245. doi:10.1080/10807039.2011.538639

Arvidsson R, Molander S, Sandén BA (2012) J Ind Ecol 16:343. doi:10.1111/j.1530-9290/2011.00429.x
Atmuri AK, Henson MA, Bhatia SR (2013) Colloid Surface A Physicochem Eng Aspect 436:325. doi:10.1016/j.colsurfa.2013.07.002
Bai C, Li Y (2012) J Contam Hydrol 136-137:43. doi:10.1016/j.jconhyd.2012.04.008
Barton LE, Auffan M, Durenkamp M, McGrath S, Bottero JY, Wiesner MR (2015) Sci Total Environ 511:535. doi:10.1016/j.scitotenv.2014.12.056
Baumann J, Köser J, Arndt D, Filser J (2014) Sci Total Environ 484:176. doi:10.1016/j.scitotenv.2014.03.023
Beer C, Foldbjerg R, Hayashi Y, Sutherland DS, Autrup H (2012) Toxicol Lett 208:286. doi:10.1016/j.toxlet.2011.11.002
Ben-Moshe T, Dror I, Berkowitz B (2010) Chemosphere 81:387. doi:10.1016/j.chemosphere.2010.07.007
Benn TM, Westerhoff PPH (2008) Environ Sci Tech 42:4133. doi:10.1021/es7032718
Berube D, Searson E, Morton T, Cummings C (2010) Nanotechnol Law Bus 7:152
Blaser SA, Scheringer M, MacLeod M, Hungerbühler K (2008) Sci Total Environ 390:396. doi:10.1016/j.scitotenv.2007.10.010
Boncagni NT, Otaegui JM, Wagner E, Curran T, Ren J, Fidalgo de Cortalezzi MM (2009) Environ Sci Tech 43:7699. doi:10.1021/es900424n
Bour A, Mouchet F, Silvestre J, Gauthier L, Pinelli E (2015) J Hazard Mater 283:764. doi:10.1016/j.jhazmat.2014.10.021
Boxall ABA, Chaudhry Q, Sinclair C, Jones A, Aitken R, Jefferson B, Watts C (2007) Current and future predicted environmental exposure to engineered nanoparticles. Tech. rep., University of York. http://randd.defra.gov.uk/Document.aspx?Document=CB01098_6270_FRP.pdf
Brunelli A, Pojana G, Callegaro S, Marcomini A (2013) J Nanopart Res 15:1. doi:10.1007/s11051-013-1684-4
Brunetti G, Donner E, Laera G, Sekine R, Scheckel KG, Khaksar M, Vasilev K, De Mastro G, Lombi E (2015) Water Res 77:72. doi:10.1016/j.watres.2015.03.003
Buha J, Mueller N, Nowack B, Ulrich A, Losert S, Wang J (2014) Environ Sci Tech 48:4765. doi:10.1021/es4047582
Buser AM, MacLeod M, Scheringer M, Mackay D, Bonnell M, Russell MH, DePinto JV, Hungerbühler K (2012) Integr Environ Assess Manag 8(4):703. doi:10.1002/ieam.1299
Comber SD, Smith R, Daldorph P, Gardner MJ, Constantino C, Ellor B (2013) Environ Sci Tech 47:9824. doi:10.1021/es401793e
Cullen E, O'Carroll DM, Yanful EK, Sleep B (2010) Adv Water Resour 33:361. doi:10.1016/j.advwatres.2009.12.001
Cundy AB, Hopkinson L, Whitby RLD (2008) Sci Total Environ 400:42. doi:10.1016/j.scitotenv.2008.07.002
Dahirel V, Jardat M (2010) Curr Opin Colloid Interface Sci 15:2. doi:10.1016/j.cocis.2009.05.006
Dale AL, Lowry GV, Casman EA (2013) Environ Sci Tech 47:12920. doi:10.1021/es402341t
Dale A, Casman EA, Lowry GV, Lead JR, Viparelli E, Baalousha MA (2015a) Environ Sci Tech 49(5):2587. doi:10.1021/es505076w
Dale AL, Lowry GV, Casman EA (2015b) Environ Sci Tech 7285–7293:49. doi:10.1021/acs.est.5b01205
David CA, Galceran J, Rey-Castro C, Puy J, Companys E, Salvador J, Monné J, Wallace R, Vakourov A (2012) J Phys Chem C 116:11758. doi:10.1021/jp301671b
Degueldre C, Aeberhard P, Kunze P, Bessho K (2009) Colloid Surface A Physicochem Eng Aspect 337:117. doi:10.1016/j.colsurfa.2008.12.2007
Doiron K, Pelletier E, Lemarchand K (2012) Aquat Toxicol 124-125:22. doi:10.1016/j.aquatox.2012.07.004
Dumont E, Johnson AC, Keller VD, Williams RJ (2015) Environ Pollut 196:341. doi:10.1016/j.envpol.2014.10.022

Dutch National Government (2015) Pollutant release and transfer registration. http://www.prtr.nl. http://www.prtr.nl
Eduok S, Martin B, Villa R, Nocker A, Jefferson B, Coulon F (2013) Ecotoxicol Environ Saf 95:1. doi:10.1016/j.ecoenv.2013.05.022
El Badawy AM, Hassan AA, Scheckel KG, Suidan MT, Tolaymat TM (2013) Environ Sci Tech 47:4039. doi:10.1021/es304580r
Elimelich M, Gregor J, Jia X, Williams R (1998) Particle deposition and aggregation: measurements, modelling and simulation. Butterworth-Heinemann, Woburn
European Commission (2011) Commission recommendation on the definition of nanomaterial. http://eur-lex.europa.eu/LexUriServ/LexUriServ.do?uri=OJ:L:2011:275:0038:0040:EN:PDF
Fabrega J, Luoma SN, Tyler CR, Galloway TS, Lead JR (2011) Environ Int 37:1. doi:10.1016/j.envint.2010.10.012
Fang J, Shan XQ, Wen B, Lin JM, Owens G (2009) Environ Pollut 157:1101. doi:10.1016/j.envpol.2008.11.006
Farmena E, Mikkelsen HN, Evensen Ø, Einset J, Heier LS, Rosseland BO, Salbu B, Tollefsen KE, Oughton DH (2012) Aquat Toxicol 108:78. doi:10.1016/j.aquatox.2011.07.007
Ferson S, Ginzburg L, Akçakaya R, Appl Biomath Rep (2001). http://www.ramas.com/whereof.pdf
Gao J, Youn S, Hovsepyan A, Llaneza VL, Wang Y, Britton G, Bonzongo J-CJ (2009) Environ Sci Tech 43:1. doi:10.1021/es803315v
Goldberg E, Scheringer M, Bucheli TD, Hungerbühler K (2014) Environ Sci Tech 48:12732. doi:10.1021/es502044k
Gottschalk F, Scholz R, Nowack B (2010a) Environ Model Software 25:320. doi:10.1016/j.envsoft.2009.08.011
Gottschalk F, Sonderer T, Scholz R, Nowack B (2010b) Environ Toxicol Chem 29:1036. doi:10.1002/etc.135
Gottschalk F, Sun T, Nowack B (2013) Environ Pollut 181:287. doi:10.1016/j.envpol.2013.06.003
Grieger KD, Fjordbøge A, Hartmann NB, Eriksson E, Bjerg PL, Baun A (2010) J Contam Hydrol 118:165. doi:10.1016/j.jconhyd.2010.07.011
Hammes J, Gallego-Urrea JA, Hassellöv M (2013) Water Res 47:5350. doi:10.1016/j.watres.2013.06.015
Hansen SF, Heggelund LR, Besora PR, Mackevica A, Boldrin A, Baun A (2016) Environ Sci Nano 3:169. doi:10.1039/C5EN00182J
He D, Bligh MW, Waite TD (2013) Environ Sci Tech 47(16):9146. doi:10.1021/es400391a
Hendren CO, Lowry M, Grieger KD, Money ES, Johnston JM, Wiesner MR, Beaulieu SM (2013a) Environ Sci Tech 47:1190. doi:10.1021/es302749u
Hendren CO, Badireddy AR, Casman E, Wiesner MR (2013b) Sci Total Environ 449:418. doi:10.1016/j.scitotenv.2013.01.078
Hotze EM, Bottero J-Y, Wiesner MR (2010) Langmuir 26:11170. doi:10.1021/la9046963
Hüffmeyer N, Klasmeier J, Matthies M (2009) Sci Total Environ 407:2296. doi:10.1016/j.scitotenv.2008.11.055
Jacobs R, van der Voet H, ter Braak C (2015) J Nanopart Res 17:251. doi:10.1007/s11051-015-2911-y
Jarvie HP, King SM (2010) Nano Today 5:248. doi:10.1016/j.nantod.2010.06.001
Kaegi R, Voegelin A, Sinnet B, Zuleeg S, Hagendorfer H, Burkhardt M, Siegrist H (2011) Environ Sci Tech 45:3902. doi:10.1021/es1041892
Kaptay G (2012) Int J Pharm 430:253
Kasel D, Bradford SA, Simunek J, Pütz T, Vereecken H, Klumpp E (2013) Environ Pollut 180:152. doi:10.1016/j.envpol.2013.05.031
Kehrein N, Berlekamp J, Klasmeier J (2015) Environ Model Software 64:1. doi:10.1016/j.envsoft.2014.10.018
Keller AA, Lazareva A (2014) Environ Sci Technol Lett 1:65. doi:10.1021/ez400106t

Koehler A, Peyer F, Salzmann C, Saner D (2011) Environ Sci Tech 45:3487. doi:10.1021/es1021763
Lattuada M, Wa H, Sefcik J, Morbidelli M (2006) J Phys Chem B 110:6574. doi:10.1021/jp056538e
Levard C, Hotze EM, Lowry GV, Brown GEJ (2012) Environ Sci Tech 46:6900. doi:10.1021/es2037405
Li K, Chen Y (2012) J Hazard Mater 209-210:264. doi:10.1016/j.jhazmat.2012.01.013
Li ZL, Sahle-Demessie E, Aly Hassan A, Sorial GA (2011) Water Res 45:4409. doi:10.1016/j.watres.2011.05.025
Liang Y, Bradford SA, Simunek J, Heggen M, Vereecken H, Klumpp E (2013) Environ Sci Tech 47(21):12229. doi:10.1021/es402046u
Liu HH, Cohen Y (2014) Environ Sci Tech 48:3281. doi:10.1021/es405132z
Liu J, Pennell KG, Hurt RH (2011a) Environ Sci Tech 45:7345. doi:10.1021/es201539s
Liu HH, Surawanvijit S, Rallo R, Orkoulas G, Cohen Y (2011b) Environ Sci Tech 45:9284. doi:10.1021/es202134p
Lorenz C, von Goetz N, Scheringer M, Wormuth M, Hungerbühler K (2011) Nanotoxicology 5:12. doi:10.3109/17435390.2010.484554
Lowry GV, Espinasse BP, Badireddy AR, Richardson CJ, Reinsch BC, Bryant LD, Bone AJ, Deonarine A, Chae S, Therezien M, Colman BP, Hsu-Kim H, Bernhardt ES, Matson CW, Wiesner MR (2012) Environ Sci Tech 46:7027. doi:10.1021/es204608d
Mackay D, Webster E, Cousins I, Cahill T, Foster K, Gouin T (2001) An introduction to multimedia models. CEMC Report 200102, Canadian Environmental Modelling Centre, Trent University, Peterborough Ontario, Canada, Canadian Environmental Modelling Centre, Trent University, Peterborough Ontario, K9J 7B8, Canada. http://www.trentu.ca/academic/aminss/envmodel/CEMC200102.pdf
Macpherson SA, Webber GB, Moreno-Atanasio R (2012) Adv Powder Tech 23:478. doi:10.1016/j.apt.2012.04.008
Mahmoodi NM, Arami M, Gharanjig K, Nourmohammadian F, Bidokhti AY (2008) Desalination 230:183
Markus A, Parsons J, Roex E, Kenter G, Laane R (2013) Sci Total Environ 456–457:154. doi:10.1016/j.scitotenv.2013.03.058
Markus A, Parsons J, Roex E, de Voogt P, Laane R (2015) Sci Total Environ 506-507:323. doi:10.1016/j.scitotenv.2014.11.056
Markus A, Parsons J, Roex E, de Voogt P, Laane R (2016) Water Res 91:214
Meesters J, Koelmans AA, Quik JT, Hendriks J, van de Meent D (2014) Environ Sci Tech 48:5726. doi:10.1021/es500548h
Mihranyan A, Strømme M (2007) Surf Sci 601:315. doi:10.1016/j.susc.2006.09.037
Mohapatra DP, Brar SK, Daghrir R, Tyagi RD, Picard P, Surampalli RY, Drogui P (2014a) Sci Total Environ 485-486:263. doi:10.1016/j.scitotenv.2014.03.089
Mohapatra D, Brar S, Daghrir R, Tyagi R, Picard P, Surampalli R (2014b) Sci Total Environ 485–486(263). doi:10.1016/j.scitotenv.2014.03.089
Müller NC, Buha J, Wang J, Ulrich A, Nowack B (2013) Evnviron Sci Process Impacts 15:251. doi:10.1039/c2em30761h
NanoRem (2015) Nanotechnology for contaminated land remediation. http://www.nanorem.eu/. http://www.nanorem.eu/
Peng Z, Doroodchi E, Evans G (2010) Powder Technol 204:91. doi:10.1016/j.powtec.2010.07.023
Peralta-Videa JR, Zhao L, Lopez-Moreno ML, de la Rosa G, Hong J, Gardea-Torresdey JL (2011) J Hazard Mater 186:1. doi:10.1016/j.jhazmat.2010.11.020
Petosa AR, Jaisi DP, Quevedo IR, Elimelech M, Tufenkji N (2010) Environ Sci Tech 44:6632. doi:10.1021/es100598h
Phenrat T, Saleh N, Sirk K, Tilton RD, Lowry GV (2007) Environ Sci Tech 41:284
Piccinno F, Gottschalk F, Seeger S, Nowack B (2012) J Nanopart Res 14:1109

Praetorius A, Scheringer M, Hungerbühler K (2012) Environ Sci Tech 46:6705. doi:10.1021/es204530n
Praetorius A, Tufenkji N, Goss KU, Scheringer M, von der Kammer F, Elimelich M (2014) Environ Sci Nano 1:317. doi:10.1039/c4en00043a
Quik JT, Vonk JA, Foss Hansen S, Baun A, van de Meent D (2011) Environ Int 37:1068. doi:10.1016/j.envint.2011.01.015
Quik J, Velzeboer I, Wouterse M, Koelmans A, van de Meent D (2014) Water Res 48:269. doi:10.1016/j.watres.2013.09.036
Quik JT, de Klein JJ, Koelmans AA (2015) Water Res 80:200. doi:10.1016/j.watres.2015.05.025
Robichaud CO, Uyar AE, Darby MR, Zucker LG, Wiesner MR (2009) Environ Sci Tech 43:4227. doi:10.1021/es8032549
Roes L, Patel MK, Worrell E, Ludwig C (2012) Sci Total Environ 417-418:76. doi:10.1016/j.scitotenv.2011.12.030
Sani-Kast N, Scheringer M, Slomberg D, Labille J, Praetorius A, Ollivier P, Hungerbühler K (2015) Sci Total Environ 49:7285. doi:10.1016/j.scitotenv.2014.12.025
Satoh A, Taneko E (2009) J Colloid Interface Sci 338:236. doi:10.1016/j.jcis.2009.06.030
Schaumann GE, Philippe A, Bundschuh M, Metreveli G, Klitzke S, Rakcheev D, Grün A, Kumahor SK, Kühn M, Baumann T, Lang F, Manz W, Schultz R, Vogel HJ (2015) Sci Total Environ 535:3. doi:10.1016/j.scitotenv.2014.10.035
Sun TY, Gottschalk F, Hungerbühler K, Nowack B (2014) Environ Pollut 185:69. doi:10.1016/j.envpol.2013.10.004
Tufenkji N, Elimelech M (2004) Environ Sci Tech 38:529. doi:10.1021/es034049r. URL http://pubs.acs.org/doi/pdfplus/10.1021/es034049r
USGS (2015) MODFLOW. http://water.usgs.gov/ogw/modflow/
Wagner S, Gondikas A, Neubauer E, Hofmann T, von der Kammer F (2014) Angew Chem Int Ed 53:12398. doi:10.1002/anie.201405050
Westerhoff P, Nowack B (2013) Acc Chem Res 46:844. doi:10.1021/ar300030n
Wikipedia (2015) DLVO Theory. https://en.wikipedia.org/wiki/DLVO_theory
Yang Y, Wang Y, Hristovski K, Westerhoff P (2015) Chemosphere 125:115. doi:10.1016/j.chemosphere.2014.12.003
Zhang H, Chen B, Banfield JF (2010) J Phys Chem C 114:14876. doi:10.1021/jp1060842
Zhang W, Yao Y, Sullivan N, Chen Y (2011) Environ Sci Tech 45:4422. doi:10.1021/es104205a. URL http://pubs.acs.org/doi/pdfplus/10.1021/es104205a
Zhou D, Keller AA (2010) Water Res 44:2948. doi:10.1016/j.watres.2010.02.025

Impact of Veterinary Pharmaceuticals on the Agricultural Environment: A Re-inspection

Paulina Łukaszewicz, Joanna Maszkowska, Ewa Mulkiewicz, Jolanta Kumirska, Piotr Stepnowski, and Magda Caban

Contents

1 Introduction 90
1.1 Antibiotics and Anti-parasitic Pharmaceuticals: Presentation and Production 91
1.2 Occurrence of VP_S in Manure and Soil 95
2 Stability of VPs in Manure and Soil 101
2.1 Biodegradation in Soil 101
2.2 Abiotic Degradation in Soil 102
2.3 Half-Lives in Manure 103
3 Leaching from Soil to Groundwater 107
3.1 Literature Data on VP Leaching 108
3.2 Occurrence of VPs in Ground Waters and Comparison to Leaching Data 111
4 VPs vs. Soil Microorganisms 113
4.1 Toxicity Toward Soil Bacteria and Fungi 114
4.2 Resistance of Bacteria 116
5 VPs vs. Plants 120
5.1 Uptake and Translocation 120
5.2 Metabolism 123
5.3 Toxicity 124
6 Current Legislation and Regulation 126
7 Conclusion and Summary 128
References 130

P. Łukaszewicz • J. Maszkowska • E. Mulkiewicz • J. Kumirska
P. Stepnowski • M. Caban (✉)
Faculty of Chemistry, Institute for Environmental and Human Health Protection, University of Gdansk, ul. Wita Stwosza 63, Gdańsk 80-308, Poland
e-mail: paulina.lukaszewicz@phdstud.ug.edu.pl; joanna.maszkowska@ug.edu.pl; ewa.mulkiewicz@ug.edu.pl; jolanta.kumirska@ug.edu.pl; piotr.stepnowski@ug.edu.pl; magda.caban@ug.edu.pl

P. de Voogt (ed.), *Reviews of Environmental Contamination and Toxicology Volume 243*, Reviews of Environmental Contamination and Toxicology 243, DOI 10.1007/398_2016_16

1 Introduction

Soil is undoubtedly one of the most important natural resources of the Earth. It is not only a habitat for many organisms, but also a platform for human activities and a supplier of many raw materials and foods (Glasener 2002). However, intensive development of the veterinary industry and agriculture, excessive fertilization and inappropriate use of pesticides is causing the entry of large amounts of pollutants into the soil (Poskrobko et al. 2007). Wastes (from wastewater treatment plants WWTP and biosolids from farmlands) are used more extensively, because the growing number of soils utilized agriculturally are now arid, and will be so even more in the future, because of climate change. These streams of wastes, raw or treated by fermentation, are rich in new classes of pollutants—i.e. pharmaceuticals, the residues of which are an emerging problem for the agricultural environment. The reasons for this will be presented below.

It is believed that residues of pharmaceuticals in the environment can pose a large problem (Andreu et al. 2009; Zuo et al. 2006; Zuo and Lin 2007). The basis for this concern is the fact that veterinary pharmaceuticals (VP_S) after administration are not completely metabolized by organisms and are excreted with urine and feces either in their native form or in the form of metabolites (Martínez-Carballo et al. 2007). Increased fertilization of farmland with natural fertilizers or sewage sludge contributes to the introduction of antibiotics into soil, surface water (via surface runoff), groundwater (through leaching), plants or other living organisms, where these pharmaceuticals can generate a number of negative consequences (Andreu et al. 2009; Beausse 2004; Chenxi et al. 2008; Ding et al. 2011; Hammesfahr et al. 2011; He et al. 2012; Hu et al. 2010a; Jacobsen et al. 2004; Kong et al. 2007; Kwon et al. 2011; Thiele-Bruhn and Peters 2007). It is believed that increased levels of antibiotics in the environment contribute, among other things, to the development of microorganisms resistant to these antibiotics (Chen et al. 2014; Kwon et al. 2011). There is also a risk that some plants are able to transfer the antibiotics from the soil into the food chain (Kong et al. 2007). Some veterinary medicines can be toxic to humans or cause allergies, thus their presence in plant food is undesirable (Lillenberg et al. 2009). VPs are quasi-persistent in the environment, because they possess degradation abilities (biotic and abiotic, dependent on compound structure and environment conditions), and are continuously entering the environment. The fate of VPs in soil is complex, including sorption, accumulation, leaching, degradation, transformation, and finally plant and organism intake. All these processes work together and can both decrease and intensify the negative effect in soil of VPs such as xenobiotics. A scheme for optimistic and worst-case scenarios resulting from VP presence in soil is shown in Fig. 1. It should be noted that some processes work in both cases, but their effects can differ. A good example of this is sorption, which can decrease toxicity, plant uptake and leaching, while also increasing the accumulation and permanence of pharmaceuticals.

VPs are one of the emerging contaminants (Campo 2015) investigated in food products, because of their potential adverse health effects in humans (Baynes et al.

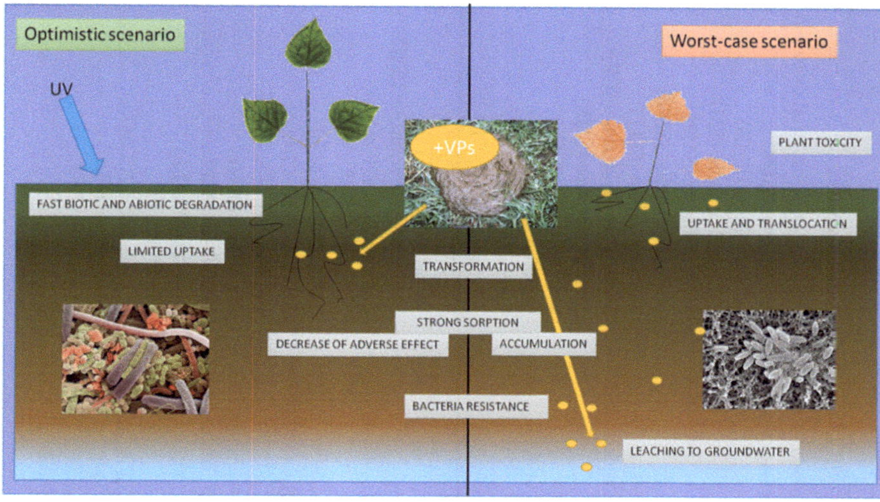

Fig. 1 The optimistic and worst-case scenarios resulting from the presence of veterinary pharmaceuticals in manure used as fertilizers

2016). In 2003 Boxall et al. posed the question "Are veterinary medicines causing environmental risk?" (Boxall et al. 2003). After 13 years, knowledge regarding this topic has increased. The influence of VPs on the soil environment has been investigated by many groups. There are ample instances in the literature concerning VPs use, sale, environmental fate, transport (sorption and leaching), stability, as well as toxicity and bacterial resistance concerning the main groups of veterinary products. Reviews of this topic are also available (Boxall et al. 2003; Kemper 2008; Sarmah et al. 2006), but new information is being reported every year. Some of these new reports show a different aspect of the presence of VP_S in natural soil and their impact on the organisms living there. The aim of this review is a re-assessment of the influence of VP_S introduced by manure fertilization on the agricultural environment.

1.1 Antibiotics and Anti-parasitic Pharmaceuticals: Presentation and Production

VPs are commonly used at therapeutic levels in livestock breeding for treating various diseases, as well as in food additives to promote animal growth. Some are also used to prevent human diseases. VPs are a broad range of compounds, including anti-parasitic drugs, anti-inflammatory medications, anesthetics, pain medications, antibiotics and specialized products used to manage reproductive, cardiovascular or metabolic conditions (Table 1). Tetracyclines (TCs),

Table 1 Most investigated VPs in environmental components

Group	Representative	Group	Representative
Antimicrobials, antibiotics (substances that kill microorganisms or suppress their multiplication or growth, treatment and prevention of bacterial diseases)			
Tetracyclines (TC_S)	Chlortetracycline (CTC) Doxycycline (DC) Oxytetracycline (OTC) Tetracycline (TC) Methacycline (MTC) Demeclocycline (DMLC) Anhydrotetracycline (AHTC) Anhydrochlortetracycline (AHCTC) Dihydrodesmycosin (DDM) 4-epi-tetracycline (4-epi-TC) 4-epi-oxytetracycline (4-epi-OTC) α-apo-oxytetracycine (α-apo-OTC) β-apo-oxytetracycline (β-apo-OTC) 2-Acetyl-2-Decarboxamido-Oxytetracycline (ADOTC)	Sulfonamides (SA_S)	Sulfadiazine (SDZ) Sulfadimidine (SMD) Sulfamethoxazole (SMX) Sulfamethizole (SMZ) Sulfadimethoxine (SDM) Sulfamethazine (SMT) Sulfachloropyridazine (SCP) Sulfaquinoxaline (SQX) Sulfathiazole (STZ) Sulfadoxine (SDX) Sulfameter (SMR) Sulfaguanidine (SGD) Sulfamerazine (SMR) Sulfacetamide (STD) Sulfamethoxypyridazine (SMXP) Sulfapyridine (SPY) Sulfisoxazole (SSX) Sulfanilamide (SFL)
Macrolides (MQ_S)	Tylosin (TYL) Azithromycin (AZM) Clarithromycin (CAM) Clindamycin (CLI) Erythromycin (ERY) Roxithromycin (ROX) Spiramycin (SPM) Vancomycin (VAN) Tilmicosin (TIL) Leucomycin (LUM) Oleandomycin (ODM)	Fluoroquinolones (FQ_S)/ quinolones (Q_S)	Ciprofloxacin (CIP) Enrofloxacin (ENR) Ofloxacin (OFN) Pefloxacin (PEF) Levofloxacin (LF) Clinafloxacin Lomefloxacin (LOM) Flumequine (FLU) Difloxacin (DFX)

β-Lactams (B-LA$_S$)	Amoxicillin (AMO) Ampicillin (APM) Cephotaxim (CTX) Cloxacilin (CLX) Cephalexin (CPX) Ceftiofur (CFT) Penicilin (PEN) Benzylpenicillin (BEN)	Aminoglycosides (AG$_S$)	Dihydrostreptomycin (DHS) Neomycin (NEO) Apramycin (AP) Paromomycin (PRM) Streptomycin (STP)
Polyether ionophores (PEI$_S$)	Salinomycin (SAL) Monensin (MON) Maduramycin (MAD) Lasalocid (LAS)	Phenicols (PQ$_S$)	Chloramphenicol (CAP) Florfenicol (FF) Thiamphenicol (TAP)
Other antibiotics	Trimethoprim (TMP) Novobiocin (NB) Bacitracin (BC) Virginiamycin (VG) Tiamulin (TAM) Nalidixic acid (NA) Narasin (NR)	Quinoxalines (QXL$_S$)	Olaquindox (OLA) Carbadox (CBX)
		Lincosamides (LINS$_S$)	Lincomycin (LIN) Clindamycin (CLIN)
Antiparasitics (BNZ$_S$), including anthelmintics (treatment of parasitic diseases, those caused by helminths, amoeba, ectoparasites, parasitic fungi, and protozoa, among others)			
Benzimidazoles (BNZ$_S$)	Albendazole (ALB) Oxfendazole (OFZ) Fenbendazole (FBZ)	Avermectins (AVR$_S$)	Doramectin (DOR) Ivermectin (IVM)
Imidazoles (IMD$_S$)/ nitroimidazoles (NID$_S$)	Metronidazole (MTZ)	Tetrahydropyrimidines (THPM$_S$)	Pyrantel pamoate (PM)
Pyrethroids (PRD$_S$)	Flumethrin (FL)	Triazines (TZ$_S$)	Toltrazuriln (TZL)

sulfonamides (SAs), macrolides (MQ_S), benzimidazoles (BNZ_S), fluoroquinolones (FQ_S) and β-lactam antibiotics (B-LA_S) belong to the group of antimicrobial and antiparasitics drugs (BNZ_S) in use for more than 50 years. These drugs have stood the test of time, because of their broad spectrum of activity, low cost and satisfactory effectiveness even in small doses. They are used to treat, for example, bacterial infections or are administered prophylactically to animals (Anderson et al. 2005; Andreu et al. 2009; Białk-Bielińska et al. 2009; Jacobsen et al. 2004; Nebot et al. 2014; Fritz and Zuo 2007). Some of these pharmaceuticals, despite the restrictions introduced in 2006 by the EU to ban all use of antibiotics in animals as growth promoters, are still being added to fodder (Dzierżawski 2012; Raich-Montiu et al. 2010).

After administration to an animal body (oral, dermal or injection), VP behavior is similar to that of human pharmaceuticals, i.e. a certain portion is metabolized, while others are eliminated unchanged. Following treatment, livestock will excrete 50–90% of the administered dose (with the parent pharmaceuticals constituting 9–30%), usually within several days. For example, SA biotransformation mainly occurs by phase I oxidation and phase II acetylation (García-Galán et al. 2008), and new types of metabolites are still under investigation (Hiba et al. 2016). Due to incomplete absorption in the animal gut and/or unmetabolized excretion [excretion rates of 30–90% of the mother compound (Sarmah et al. 2006)], the substances can enter the environment via the use of manure as a soil fertilizer, as well as via the anaerobic fermentation process in biogas plants when manure is used as a substrate (the elimination rate differs between compounds and conditions).

Annual growth rates (%) of total livestock production are set to increase, especially in developing countries (Steinfeld 2004). The Asian pharmaceutical market is growing at a rate of 10–15% annually, compared to a 5–7% growth in G7 countries (Rehman et al. 2013). The Food and Agriculture Organization of the United Nations (FAO) indicates that the value of global aquaculture production increased from approximately $50 billion to nearly $140 billion in the years between 2003 and 2012 (Lillicrap et al. 2015). In 2013, 1452 tons of VPs were supplied with TC_S and penicillins (PEN_S) being the most distributed active compounds (454 tons and 473 tons, respectively), followed by SA_S (152 tons) (Spielmeyer et al. 2015). Almost 100,000 different chemicals are registered in the EU, of which 30,000 are sold in quantities above 1 ton (Beausse 2004). It is estimated that more than 20,000 tons of antibiotics are produced annually, of which more than half are used in veterinary medicine (Li et al. 2011; O'Connor and Aga 2007). According to a report of the American Association of Concerned Scientists, 11.2–12.8 million kilograms of antibacterial substances were used on farm animals for medicinal purposes in 2001 alone (Dzierżawski 2012). In 2004, pharmaceutical wholesalers in Denmark sold 114 tons of veterinary drugs (not counting those available without a prescription), Finland reported 13.3 tons, Sweden-16.1 tons, while the amount of sales of veterinary medicines in Germany and France stood at respectively 668.8 and 1270 tons (Kools et al. 2008).

In Europe, the most widely marketed antibacterial agents are TC_S, followed by PEN_S, SA_S, MQ_S, FQ_S and third-and fourth-generation cephalosporins (CLS_S), but patterns vary between countries (Grave et al. 2014). In 2012 in Poland, it was established that the total sales of antibacterial veterinary medicinal products amounted to 519 tons, of which the share of TC_S, PEN_S and SA_S was, respectively, 41, 25, and 10% (Ministry of Agriculture and Rural Development 2013).

All these facts indicated that the amount of VPs which reach the environment is substantial and will increase every year. Although VPs do not persist in the environment (their stability depends on several factors), their continuous supply to natural matrices makes them semi-persistent pollutants. The concentration of VP_S already present in the environment (water system and soils) is already worrying.

1.2 Occurrence of VP_S in Manure and Soil

The development of the veterinary industry and the intensification of agriculture are the main causes of the presence of veterinary antibiotics in almost every component of the environment (Dghrir and Drogui 2013; Homem and Santos 2011; Maia et al. 2014; Solliec et al. 2016; Thiele-Bruhn 2003). While there is a significant amount of literature data on the occurrence of pharmaceuticals in surface waters, knowledge of the solid components, particularly including soil, is limited (Andreu et al. 2009). It is a disturbing fact that every year considerable amounts of bio-fertilizers (e.g. manure and sludge) are applied to fields. In 2006 alone, the United States produced about 8 million tons of sludge, of which 50% was introduced into the soil (Chenxi et al. 2008). In contrast, five times the amount of manure—41 million tons—was generated in Korea, of which almost 81% was used as compost or manure on croplands (Kwon et al. 2011). In the EU, about 37% of the 6.6 million tons of sewage sludge is used annually for soil fertilization (Jjemba 2002). Via the fertilization of farmland with manure, a lot of contaminants, including VPs, are introduced into the soil. The amount of antibiotics entering fields in fertilizers are estimated at kilograms per hectare (Kemper 2008; Li et al. 2011; Thiele-Bruhn 2003). The amount of antibiotics contained in manure, which is considered the main source of these substances (Beausse 2004; Kummerer 2003; Li et al. 2011), may remain constant or increase as a result of the re-transformation of metabolites to their native form (Kemper et al. 2008). An example of this are metabolites in the form of conjugates (acetylated metabolites). These substances show no activity, but after entry into fertilizer the acetyl group may detach, which will cause the release of the active substance (Christian et al. 2003; Kemper et al. 2008). Furthermore, transformation products may have different physico-chemical properties than the parent compound and can exhibit greater toxicity in the environment and higher mobility (García-Galán et al. 2009; Mitchell et al. 2014; Solliec et al. 2016). Dust escaping from stables or barns and being deposited on the surface of the ground, or wastewater discharges from veterinary clinics are considered to be further sources of veterinary pharmaceuticals (Kemper 2008; Le-Minh et al. 2010; Masse et al. 2014; Sarmah et al. 2006; Thiele-Bruhn 2003; Chen et al. 2013).

An increasing number of reports associated with the presence of pharmaceuticals in the environment are appearing in the literature. It is believed that the most frequently detected veterinary antibiotic groups in fertilizers and soils are TC_S, SA_S, FQ_S, MQ_S and PEN_S (Beausse 2004; Picó and Andreu 2007; Tadeo et al. 2012). Information on the occurrence of anthelmintics is very limited (Horvat et al. 2012). Examples of detected concentrations of veterinary drugs in various environmental samples are presented in Table 2.

Generally, the concentration levels of veterinary compounds in solid components are in the range of μg kg^{-1}–mg kg^{-1}. However, the determined quantities are significantly lower in soils (ppb range), probably due to degradation and leaching or water uptake (Hammesfahr et al. 2011; Hu et al. 2010b; Masse et al. 2014; Rabølle and Spliid 2000; Ramaswamy et al. 2010a; Wu et al. 2010a, b). The following review of the literature (Table 2) confirms that the most frequently detected and determined compounds are TC_S, which is undoubtedly related to their leading position in terms of sales and consumption. In contrast, much less information can be found on the presence of VPs from the MQ_S group or LCS_S. Their concentrations in both manure and soil are usually at μg kg^{-1}. Pan et al. described the frequency of appearance of the most widely used group of antibiotics in samples of pig manure. According to the results, TCs are found in 85–97% of samples, SAs in 52%, while MQs in only 5% (Pan et al. 2011). TCs and SAs are commonly determined in soils and fertilizers. It is stated that the concentration of these pharmaceuticals in pig manure is 4 mg kg^{-1} for TCs and 3.5 mg kg^{-1} for SAs (Thiele-Bruhn 2003). In agricultural soils in Denmark, the concentration of OTC is in the range of 2.5–50.0 μg g^{-1}, while in pig manure the amount of this compound is much greater: 33–2000 μg g^{-1} (Kong et al. 2007). In Germany, the concentration of TCs is 450–900 μg kg^{-1} (Sarmah et al. 2006). In Turkey, the maximum concentration of OTC was reported to be 500 μg kg^{-1} in fertilized agricultural soils (Li et al. 2011). Examining the residues of TCs in 100 Spanish soil samples, Andreu et al. noted OTC, TC and CTC in 25 samples, and the most commonly detected pharmaceutical was OTC (15.7–105.4 μg kg^{-1}) (Andreu et al. 2009). In a study conducted by Hu et al., the most commonly determined group of veterinary medicines in Chinese agricultural soils were FQ_S (19–120 μg kg^{-1}). The concentrations of other pharmaceuticals were 12–46 μg kg^{-1} for TCs and 2–9 μg kg^{-1} for SAs (Hu et al. 2012). Higher concentrations of SA_S are recorded in the work by Garcia-Galan et al., wherein SMT and STZ are determined at the levels of 8.7 and 12.4 mg kg^{-1}, respectively (García-Galán et al. 2009). On the other hand, the number of MQs measured in these soils has been established at the level of 13–67 μg kg^{-1} (Thiele-Bruhn 2003).

The detected amount of the compounds largely depends on the sorption capacity of the VPs, the type of fertilizer used and the depth from which soil samples were collected. In the case of natural fertilizers (pig and poultry), it is estimated that the concentrations of the most commonly used antibiotics are up to 216 mg L^{-1} (Masse et al. 2014), where according to Karci et al. higher amounts of antimicrobial agents are detected in samples of fresh poultry manure (Karci and Balcioğlu 2009). Some literature data show that VPs (particularly TCs) can be found up to 40 cm below the

Table 2 Occurrence of VPs in soils, manure and biosolids (Andreu et al. 2009; Boxall et al. 2006; Campagnolo et al. 2002; Christian et al. 2003; Ding et al. 2011; García-Galán et al. 2009; Haller et al. 2002; Hammesfahr et al. 2011; Hamscher et al. 2002; Hu et al. 2010a, 2010b, 2012; Jacobsen et al. 2004; Kong et al. 2007; Li et al. 2011; Martínez-Carballo et al. 2007; Qiao et al. 2012; Sarmah et al. 2006; Shelver et al. 2010; Thiele-Bruhn 2003; Uslu et al. 2007; Winckler et al. 2003; Zhou et al. 2012)

Group of veterinary pharmaceuticals	Matrix	Determined Concentration	References
TCs			
TC OTC CTC	Animal manure	0.36–23 mg kg^{-1} 0.21–29 mg kg^{-1} 0.10–46 mg kg^{-1}	(Martínez-Carballo et al. 2007)
OTC CTC		1.6–136 mg kg^{-1} 1.1–26 mg kg^{-1}	(Winckler et al. 2003)
CTC OTC TC DC MTC		35,500 ng g^{-1} 661 ng g^{-1} 821 ng g^{-1} 296 ng g^{-1} 4.02 ng g^{-1}	(Zhou et al. 2012)
CTC TC OTC		68.0–1000 μg L^{-1} 25–410 μg L^{-1} 25–410 μg L^{-1}	(Campagnolo et al. 2002)
OTC		33–2000 mg g^{-1}	(Kong et al. 2007)
OTC TC CTC		5.3–183.5 mg kg^{-1} 8.3–43.5 mg kg^{-1} 0.4–26.8 mg kg^{-1}	(Hu et al. 2010b)
CTC TC		0.1 mg kg^{-1} 4.0 mg kg^{-1}	(Hamscher et al. 2002)
TC OTC CTC DC MTC		15,264 μg kg^{-1}	(Qiao et al. 2012)
CTC TC		0.15–14.7 mg kg^{-1} 0.18–0.84 mg kg^{-1}	(Hu et al. 2010a)
TC CTC		1.1 mg kg^{-1} 1.7 mg kg^{-1}	(Martínez-Carballo et al. 2007)
OTC CTC TC DC	Soil	15.7–105.4 μg kg^{-1} 5.8–34.4 μg kg^{-1} 18.8–64.3 μg kg^{-1} 12.1–45.7 μg kg^{-1}	(Andreu et al. 2009)
TC CTC		86.2–198.7 μg kg^{-1} 4.6–7.3 μg kg^{-1}	(Hamscher et al. 2002)
CTC		10–15 μg kg^{-1}	(Jacobsen et al. 2004)
OTC		305 μg kg^{-1}	(Boxall et al. 2006a, b, c)
OTC TC CTC		124–2683 μg kg^{-1} 20.9–105 μg kg^{-1} 33.1–1079 μg kg^{-1}	(Hu et al. 2010b)

(continued)

Table 2 (continued)

Group of veterinary pharmaceuticals	Matrix	Determined Concentration	References
TC CTC OTC		12–46 μg kg^{-1}	(Hu et al. 2012)
OTC		2.5–50 μg g^{-1}	(Kong et al. 2007)
TC		450–900 μg kg^{-1}	(Thiele-Bruhn 2003) (Sarmah et al. 2006)
OTC TC CTC		79.7 μg kg^{-1} 74.4 μg kg^{-1} 104.6 μg kg^{-1}	(Li et al. 2011)
CTC DMLC DC OTC TC	Treated sewage sludge	69.6–346.6 μg kg^{-1} 36.6–131.2 μg kg^{-1} 149.6–291.2 μg kg^{-1} 51.9–743.6 μg kg^{-1} 281.9 μg kg^{-1}	(Ding et al. 2011)
SA_S			
SMX SDX SCP	Animal manure	2.0–5.7 mg kg^{-1} 0.1–32.7 mg kg^{-1} 0.3–2.4 mg kg^{-1}	(Hu et al. 2010b)
SMD		1 mg kg^{-1}	(Christian et al. 2003)
SMR TMP		47 ng g^{-1} 6.26 ng g^{-1}	(Zhou et al. 2012)
SGD SDZ SMD SMX SDM		20 mg kg^{-1}	(Haller et al. 2002)
SMT STZ		8.7 mg kg^{-1} 12.4 mg kg^{-1}	(García-Galán et al. 2009)
SMT SDM TMP		2.5–400 μg L^{-1} 2.5 μg L^{-1} 2.5 μg L^{-1}	(Campagnolo et al. 2002)
SMT STZ		2250–5060 ng L^{-1} 785–1700 ng L^{-1}	(Shelver et al. 2010)
SDM SDZ TMP		20 mg kg^{-1} 51–91 mg kg^{-1} 17 mg kg^{-1}	(Martínez-Carballo et al. 2007)
SMD	Soil	15 μg kg^{-1}	(Christian et al. 2003)
SA_S		90 μg kg^{-1}	(Hammesfahr et al. 2011)
SMX SDX SCP		0.1–0.9 μg kg^{-1} 1.2–91 μg kg^{-1} 1.3–2.5 μg kg^{-1}	(Hu et al. 2010b)
SDZ TMP		0.8 μg kg^{-1} 0.5 μg kg^{-1}	(Boxall et al. 2006a, b, c)

(continued)

Table 2 (continued)

Group of veterinary pharmaceuticals	Matrix	Determined Concentration	References
STD SCP SDM SMR SMD SMX STZ SMZ SMXP SPY SSX		2–9 μg kg^{-1}	(Hu et al. 2012)
SMR SMT SMR SMX SDM		114.8–321.4 μg kg^{-1}	(Li et al. 2011)
SDZ SMR SMT SMX	Treated sewage sludge	562 μg kg^{-1} 112–668.9 μg kg^{-1} 127.8–131.8 μg kg^{-1} 4.8–35.9 μg kg^{-1}	(Ding et al. 2011)
FQs			
CIP	Animal manure	0.1 mg kg^{-1}	(Hu et al. 2010b)
ENR OFN		16.5 ng g^{-1} 8.15 ng g^{-1}	(Zhou et al. 2012)
OFN CIP PEF		1.2–15.7 mg kg^{-1} 0.1–4.3 mg kg^{-1} 3.3–24.7 mg kg^{-1}	(Hu et al. 2010b)
ENR		2.8–8.3 mg kg^{-1}	(Martínez-Carballo et al. 2007)
CIP NOR ENR LOM	Soil	13.7–1347.6 μg kg^{-1}	(Li et al. 2011)
FQs		6–52 μg kg^{-1}	(Thiele-Bruhn 2003)
NOR CIP ENR OFN		19–120 μg kg^{-1}	(Hu et al. 2012)
OFN CIP		0.1 μg kg^{-1} 10.3–30.1 μg kg^{-1}	(Hu et al. 2010b)
CIP		0.37 mg kg^{-1}	(Martínez-Carballo et al. 2007)
NOR CIP		0.29–0.40 mg kg^{-1}	(Golet et al. 2002)
ENR		0.013–0.204 mg kg^{-1}	(Uslu et al. 2007)

(continued)

Table 2 (continued)

Group of veterinary pharmaceuticals	Matrix	Determined Concentration	References
MQ_S and LCS_S			
LUM	Animal manure	15.2 ng g^{-1}	(Zhou et al. 2012)
ERY LIN		2.5 μg L^{-1} 2.5–240 μg L^{-1}	(Campagnolo et al. 2002)
LIN		0.12–3.8 mg kg^{-1}	(Hu et al. 2010b)
TYL		<110 μg kg^{-1}	(Jacobsen et al. 2004)
LIN	Soil	8.5 μg kg^{-1}	(Boxall et al. 2006a, b, c)
TYL A		10–20 μg kg^{-1}	(Jacobsen et al. 2004)
LIN		1.1–11.7 μg kg^{-1}	(Hu et al. 2010b)
MQ_S		13–67 μg kg^{-1}	(Thiele-Bruhn 2003)
ERY LIN	Treated sewage sludge	10.4–62.8 μg kg^{-1} 2.6–8.7 μg kg^{-1}	(Ding et al. 2011)

surface of soil fertilized with animal slurry. There are no signs of the elution of these substances to deeper levels (Andreu et al. 2009). For example, in experiments conducted in Germany the presence of TC and CTC in slurry was reported in amounts of 4 and 1 mg kg^{-1}. The concentration of these pharmaceutical in soil samples averaged 86.2 μg kg^{-1} in the 0–10 cm layer and 171.7 μg kg^{-1} at 20–30 cm (Sarmah et al. 2006). Zhang et al. found TCs in a layer of soil (0–20 cm) in concentrations of 3.9–43.4 μg kg^{-1}, while at a depth of 20–30 cm the concentration of these compounds was 2.8 μg kg^{-1} (Zheng et al. 2012). Other assays confirmed the presence of antibiotics in sandy soils fertilized with slurry, where OTC, TC, CTC and SMT were detected at maximum concentrations equal to, respectively, 27, 443, 93 and 4.5 μg kg^{-1} in the 0–30 cm layer (Sarmah et al. 2006). Other sources indicate that SMT is detected only in the top layer of the soil (15 cm) (Shelver et al. 2010).

Some antibacterial agents (above all, TCs and FQ_S) may remain in soils for many years (Beausse 2004; Chen et al. 2014; Díaz-Cruz et al. 2003; Jechalke et al. 2014; Karci and Balcioğlu 2009; Kemper 2008; Masse et al. 2014; Rabølle and Spliid 2000), and this is associated with their strong sorption on the surface of the soil (Picó and Andreu 2007; Wang and Yates 2008). For example, Tamtam et al. describe the presence of veterinary antibiotics in agricultural soils near Paris after 4 years of long-term wastewater irrigation (Chen et al. 2014; Tamtam et al. 2011), while SAs have been determined in soils after a year of fertilization (Shelver et al. 2010). Most antibiotics are adsorbed very quickly. When the frequency of soil fertilization with manure exceeds the degree of degradation of the antibiotics, they can be expected to accumulate in soils (Białk-Bielińska et al. 2011; García-Galán et al. 2009; Kemper 2008; Qiao et al. 2012). The above research results show that VPs are a major environmental problem worldwide, and that their detection in sediments, soils and waters raises concern about the possible adverse effects of these compounds on natural ecosystems (Andreu et al. 2009). However, more research is needed to understand the behavior of VPs in soils and fertilizers (Masse et al. 2014) and to assess their potential risk to human health (Liu et al. 2014).

2 Stability of VPs in Manure and Soil

The large consumption of a veterinary antibiotic results in its presence in the environment. Despite the fact that pharmaceuticals in manure or soil occur in low concentrations (ppb–ppm range) and can undergo processes of degradation, their continuous introduction into the ecosystem may be dangerous (Białk-Bielińska et al. 2011; García-Galán et al. 2009; Homem and Santos 2011; Jjemba 2002; Schlusener and Bester 2006). Both biotic and abiotic processes determine the environmental fate of pharmaceuticals (Montesdeoca-Esponda et al. 2012), as is discussed in further sections. Abiotic degradation processes of VPs are dependent on pH, redox conditions and exposure to light.

2.1 *Biodegradation in Soil*

It is believed that the biodegradation of chemicals is probably one of the most important environmental processes, and where a greater amount of manure is placed on fields, its intensity is greater (Kemper 2008; Sittig et al. 2014). During biodegradation, a chemical substance may be mineralized or transformed into a decomposition product (Barra Caracciolo et al. 2015; Ingerslev and Halling-Sørensen 2001).

It is believed that the amount of organic matter in soils can both increase and reduce the rate of biodegradation (Ingerslev and Halling-Sørensen 2001). In the case of FQ_S and TCs, which strongly bind to the surface of the soil, it is suspected that the sorption process can delay biotic and abiotic degradation, which is why they are persistent in sediments, soils and manure (Lillenberg et al. 2009; Uslu et al. 2007).

SAs, FQ_S and TMP are considered to be compounds which are resistant to microbial degradation (Baran et al. 2006; Białk-Bielińska et al. 2011; Le-Minh et al. 2010; Montesdeoca-Esponda et al. 2012; Sittig et al. 2014; Van Doorslaer et al. 2014). However, various data can be found in the literature concerning their durability. For example, the constant biodegradation rate of ciprofloxacin in soil is 0.39 d^{-1}. The half-life period of this compound is longer (19 days) in sterilized soil than in non-sterilized soil (9 days) (Zhang et al. 2012). Nevertheless, it has also been proven that this compound does not degrade under anaerobic conditions (Chenxi et al. 2008; Thiele-Bruhn 2003). In the case of SAs (SMD), which do not bind too strongly to the surface of the soil, there is information about their stability in both the slurry and in the soil (Christian et al. 2003). The persistence of VPs from the TC_S group, particularly CTC, in Danish soils is established at 25–34 days

(Daghrir and Drogui 2013). According to reports in the literature, VPs from the group of MQs and AG_S decompose well in the environment (Haller et al. 2002). MQ_S (especially TYL) are stable in neutral conditions, but under acidic conditions TYL A degrades to TYL B (Sarmah et al. 2006; Thiele-Bruhn 2003). The half-life of this compound in sandy soil is 49–67 days (Sarmah et al. 2006). Different values were presented by Schlusener et al. where, during a 120-day experiment of the oxygen degradation of five MQ_S antibiotics in soil (TYL, ERY, ODM, SAL and TAM), it was shown that the half-life of these compounds is 8, 20, 27, 16, and 5 days, respectively (Schlusener and Bester 2006). B-LA_S antibiotics are hydrolytically and biologically degradable in a matter of hours or days depending on the soil moisture. The rate of degradation of these compounds has been assessed as 5–67 days (Carlson and Mabury 2006; Halling-Sørensen et al. 2005; Jechalke et al. 2014). This is due to the structure of B-LA_S antibiotics (such as PEN, BEN or CLX), consisting of a β-lactam ring, which contributes to the rapid degradation of these compounds in the environment (Kemper et al. 2008).

2.2 *Abiotic Degradation in Soil*

Among many processes, photodegradation contributes to the decomposition of pharmaceuticals on the surface of soils (Jechalke et al. 2014; Chen et al. 2012). Fertilizers spread on soils contain pharmaceuticals that are exposed to sunlight, but the process of photolysis may be significant only when the pharmaceuticals are located in a thin layer of distributed manure. SAs, TCs or FQ_S tend to undergo photolysis in aqueous solutions, but information on the degradation of these compounds on the surface of the soils is not only limited but also inconsistent (Martínez-Carballo et al. 2007; Thiele-Bruhn and Peters 2007).

It is believed, however, that in the case of FQ_S the rate of photodegradation in soil is lower than in water, which may be related to the presence of humic acids (Picó and Andreu 2007). However, according to Beausse et al., the photodegradation of FQs in soils is incomplete, which results in the detection of these compounds in soil samples—0.3 mg kg^{-1}, even 8–21 months after fertilization (Beausse 2004; Jechalke et al. 2014; Picó and Andreu 2007). Similar observations were made by Kemper, who believes that the effect of light is negligible when these compounds are located in a sediment or suspension (Kemper 2008). The photolytic and chemical stability of FQs is conditioned by the presence and durability of the quinolone ring (Babić et al. 2013). It is known that in aqueous conditions degradation products of TCs are formed by epimerisation, dehydration or proton transfer. In pH 2–6 4-epi-TCs can be created, which in special alkaline conditions revert to an active form (Halling-Sørensen et al. 2003a, b; Jia et al. 2009; Solliec et al. 2016). In acidic conditions ($pH < 2$) AHTC and AHCTC are created, of which only AHTC is stable. AHCTC is transformed to α and β-apo-OTC forms (Halling-Sørensen et al. 2003a, b; Jia et al. 2009; Solliec et al. 2016; Wu et al.

2011). However, literature data on products of transformation in solid environments (especially in soils) are limited.

The durability of antibiotics in soils is highly dependent not only on their susceptibility to degradation, but also on leaching in water, the structure of the compound, the physico-chemical properties and type of soil, and the climatic conditions, in particular the temperature (Boxall et al. 2004; Díaz-Cruz et al. 2003; Horvat et al. 2012; Picó and Andreu 2007), which accelerates the decomposition of compounds. For example, studies on the stability of CTC in soil after 30 days of incubation have shown the presence of 44% of the starting concentration of the compound at 30 °C, 88% at 20 °C and the lack of any degradation of CTC at 4 °C (Hamscher et al. 2002). It has also been proven that the time of the use of manure on the field can affect the degradation rate of compounds. It has been shown that, in soil fertilized with manure, the half-life of IVM in winter ranges from 91 to 217 days, while in the summer it is much shorter (7–14 days) (Boxall et al. 2004).

2.3 *Half-Lives in Manure*

Half-lives of many antibiotics in solid components often vary from a few days to several weeks (Díaz-Cruz 2007; Masse et al. 2014), but their degradation in manure, soil or sediment proceeds according to the same mechanisms. Boxall et al. presented a summary of the half-lives of veterinary antibiotics in manure (Boxall et al. 2004). According to these data, the most persistent groups of pharmaceuticals are TC_S and Q_S ($t_{1/2}$ > 100 days). In the case of SA_S, AG_S and MQ_S their half-lives are respectively <8–30 days, 30 days, <2–21 days. The least durable compounds are pharmaceuticals from the group of $B\text{-}LA_S$; their half-life is only 5 days. Other sources say that AVR_S (belonging to the MQ_S) remain in manure for a long time (Raich-Montiu et al. 2011). However, according to Wang et al. the half-life of TYL in the feces of animals (cattle, chicken and swine) is 6.2, <6.2 and 7.6 days, respectively (Teeter and Meyerhoff 2003; Wang and Yates 2008).

The degree of VP degradation in manure is connected with aerobic conditions; however, manure can also be composted (anaerobic conditions). In the case of TYL, $t_{1/2}$ in veal manure is <2 days (under methanogenic conditions) (Liguoro et al. 2003; Loke et al. 2000), so degradation is faster than in soil. After 8 months of incubation in conventional-lagoon slurry (22 °C, anaerobic conditions), Kolz et al. reported the presence of degradation products of TYL-primarily TYL B, D and DDM (Kolz et al. 2005), which proves the incomplete degradation of TYL in lagoons. Moreover, other sources indicate that, in the case of fertilizers stored in lagoons, concentrations of antibiotics can be stable for up to 150 days (Kuchta and Cessna 2009; Ramaswamy et al. 2010b).

In the case of TC_S, it is considered that they are more rapidly decomposed in liquid porcine manure under aerobic than anaerobic conditions (Thiele-Bruhn 2003; Wang and Yates 2008). During laboratory tests on the stability of TC in liquid manure, a significant reduction of the concentration and the creation of an optical isomer of 4-epi-TC was reported (Sarmah et al. 2006). The same study also observed that the degradation of TC_S in the fertilizer was completed more quickly because of the higher pH. O'Connor et al. showed that after 6 days 29–64% TCs were found in manure, while no degradation products were identified (O'Connor and Aga 2007). On the other hand, Storteboom showed that OTC remains longer in dairy manure (9.8d or 17.7d-depending on intensity management) than in horse manure (Masse et al. 2014; Storteboom et al. 2007). In the case of OTC, it is estimated that the half-life of this compound in calf fertilizer is 30 days; however, this compound is detected after 5 months (820 μg kg^{-1}) (Andreu et al. 2009; Liguoro et al. 2003). According to Halling-Sørensen (2001), the half-life of MTZ, OLA and TYL in soil-manure slurries is respectively 9.7–26.9 days, 5.8–8.7 days and 3.3–8.1 days (Ingerslev and Halling-Sørensen 2001; Li 2014). A similar half-life (8.3 days) was recorded by Schlusener and Bester (2006). According to Wang et al., OTC degrades faster (during laboratory aerobic incubation) in manure at a temperature of 35–45 °C than at 15–25 °C (Wang and Yates 2008); after 64 days 59% was degraded (anaerobic digestion); also products of the degradation of 4-epi-OTC α-apo-OTC, and β-apo-OTC were detected (Wang and Yates 2008). It is believed that up to 75% of TCs (OTC) are removed at during anaerobic digestion (Masse et al. 2014), although information on this topic is still limited (Arikan et al. 2006). The significant deviations in half-life times of VPs degradation are connected with the variability of chemical structures and properties in this group of pharmaceuticals.

According to reports in the literature, composting can significantly reduce levels of veterinary antibiotics (Zhang et al. 2013a) (Liguoro et al. 2003). Composting is a controlled aerobic process, in which diverse groups of microorganisms degrade organic matter. In the composting process, temperatures often exceed 40 °C as a result of great biological activity (Dolliver et al. 2008). In the manure of calves after 35 days of composting, OTC concentration decreased by 95% (Arikan et al. 2007). A reduction in concentration was also observed in the case of CTC in poultry manure—90% after 42 days of composting, while in pig manure only 27% of the compound was reduced (Ho et al. 2013). Dolliver et al. reported a 99% decrease in the concentration of CTC in turkey manure over 35 days (Dolliver et al. 2008; Ho et al. 2013). Speltini et al. provided information on the durability of CIP during composting, suggesting it was greater than the durability of other compounds such as SDZ and CTC (Speltini et al. 2015). In the case of MON and TYL, after 35 days of composting in turkey manure concentrations of these compounds decreased by 54 and 76%, respectively (Ho et al. 2013), but no SMT degradation was observed during 35 days of turkey manure composting (Ho et al. 2013). Nevertheless, the use of manure after composting as a fertilizer is safer for the soil (Ho et al. 2013; Ramaswamy et al. 2010b).

An example summary of the half-lives of selected veterinary medicines in various environmental matrices has been presented in Table 3. The data are diverse, so the comparison of results and the drawing of clear conclusions is difficult. However, it can be observed that the degradation of drugs occurs faster in aerobic

Table 3 Half-lives of selected veterinary medicines in various environmental matrices (Angenent et al. 2008; Arikan et al. 2006; Boxall et al. 2004; Carlson and Mabury 2006; Dolliver et al. 2008; Halling-Sørensen et al. 2003a, b, 2005; Ho et al. 2013; Ingerslev and Halling-Sørensen 2001; Ingerslev et al. 2001; Kay et al. 2004; Liguoro et al. 2003; Loke et al. 2000; Pan and Chu 2016a; Schlusener and Bester 2006; Storteboom et al. 2007; Teeter and Meyerhoff 2003; Wang and Yates 2008; Winckler and Grafe 2001; Zhang et al. 2012)

Process	Compound	Half-life (days)	Ref.
Soil			
Degradation in sandy loam soil (aerobic conditions)	ERY ROX SAL TAM ODM TYL	20 >120 5.0 16 27 8.3	(Schlusener and Bester 2006)
Degradation in soil (aerobic and anaerobic conditions)	CIP	19.1 (anaerobic) 9.2 (aerobic)	(Zhang et al. 2012)
Degradation in soil	OTC SCP	18.2 3.5	(Kay et al. 2004)
Degradation in soil	TYL CTC	49–67 25–34	(Halling-Sørensen et al. 2005)
Degradation in soil (aerobic and anaerobic conditions)	TC	31.5–57.8 (aerobic) 43.3–86.6 (anaerobic)	(Pan and Chu 2016a)
	SMT	24.8–49.5 (aerobic) 34.7–57.8 (anaerobic)	
	NOR	2.91–40.8 (aerobic) 5.60–53.4 (anaerobic)	
	ERY	6.4–40.8 (aerobic) 11.0–57.8 (anaerobic)	
	CAP	6.70–43.3 (aerobic) 8.60–53.3 (anaerobic)	
Degradation in soil (aerobic conditions)	OTC	56 Days in non-amended soil	(Wang and Yates 2008)
Manure			
Degradation in broiler manure compost (aerobic conditions)	DC ENR ERY FLU NOR SDZ TIL TMP TYL	3.8 2.8 1.4 1.3 2.1 1.4 2.0 3.7 2.2	(Ho et al. 2013)
Degradation in cattle, chicken and swine manure (aerobic conditions)	TYL	6.2 (cattle) <7.6 (chicken) 7.6 (swine)	(Teeter and Meyerhoff 2003)

(continued)

Table 3 (continued)

Process	Compound	Half-life (days)	Ref.
Degradation in manure-amended and manure free	TYL CTC MON	4.5–6.1 21–24 3.3–3.8	(Carlson and Mabury 2006)
Degradation in pig slurry	OTC SCP	79 127	(Kay et al. 2004)
Degradation in spiked turkey litter during composting (aerobic conditions)	CTC MON SMT TYL	1 17 No biodegradation 19	(Dolliver et al. 2008)
Degradation in pig manure	TC	55–105	(Winckler and Grafe 2001)
Degradation in Bedding + pig manure	OTC	30	(Liguoro et al. 2003)
Degradation in dairy manure, horse manure, feedlot manure (aerobic conditions)	TYL MON CTC OTC TC	Horse manure: 4.2–9.8 Horse manure: 14.7–30.1 Horse manure: 5.1–8.4 Dairy manure: 5.8–6.8 Feedlot manure: 13.4–13.5 Dairy manure: 9.8–17.7 Feedlot manure: 15.2–31.1 Feedlot manure: 6.5–17.2	(Storteboom et al. 2007)
Degradation in manure (anaerobic conditions)	OTC	56	(Arikan et al. 2006)
Degradation in swine waste (anaerobic conditions)	TYL	2.49 h	(Angenent et al. 2008)
Degradation in animal manure (aerobic conditions)	OTC	8.1	(Wang and Yates 2008)
Degradation in aqueous phase in manure	TYL	<2	(Loke et al. 2000)
Others			
Degradation in soil-manure slurries (aerobic conditions)	MTZ OLA TYL	13.1–26.9 5.8–8.8 3.3–8.1	(Ingerslev and Halling-Sørensen 2001)
Degradation in sediment slurry (aerobic and anaerobic conditions)	OTC OLA MTZ TYL	42–46 (aerobic) 3–8 (aerobic), 21.5 days (anaerobic) 14–104 (aerobic), 3–75 (anaerobic) 9.5–40 (aerobic)	(Ingerslev et al. 2001)
Degradation in soil/feces mixtures (summer and winter)	IVM	7–14 (summer) 91–217 (winter)	(Boxall et al. 2004)
Degradation in soil interstitial water (anaerobic conditions)	OTC α-apo-OTC β-apo-OTC 4-epi-OTC ADOTC	38.6 (light), 29.6 (dark) 2.3 (light), 15.4 (dark) 270 (light), 90.5 (dark) 142.4 (light), 29.7 (dark) 41.7 (light), 38.2 (dark)	(Halling-Sørensen et al. 2003a, b)
Degradation in manure-amended soil (aerobic conditions)	OTC	33	(Wang and Yates 2008)

conditions. The fastest degradable group of drugs (among those presented) in soil and manure are MQ_S. Additionally, one can claim that composting is an effective process for eliminating the residues of veterinary antibiotics from manure.

3 Leaching from Soil to Groundwater

Sorption to soil can be considered to be an elimination process; thus, it affects the bioavailability of pollutants in the environment. It is, therefore, an immensely important process in terms of its potential hazardous effect on organisms living in both soil and water compartments. This phenomenon is based on various interactions of sorbates with soil particles including weak (e.g. van der Waals, electrostatic) or strong interactions (e.g. covalent bonding). Depending on the reversibility of the sorption process, we can distinguish equilibrium sorption (reversible, pollutant considered bioavailable), and kinetic (entrapped) adsorption known as sequestration (temporarily decreased bioavailability due to diffusion into micro-and nanopores). If sequestered residues are released, they can become a bioavailable fraction, whereas the irreversible parts of sequestered pollutants are considered non-extractable residues and therefore not bioavailable (Jechalke et al. 2014). Although both physicochemical properties of chemicals, and soil properties are crucial in the evaluation of sorption mechanisms, an assessment of the fate of chemical compounds in soil environments should also take into account the methodological approach. Therefore, in this review the authors focus on a comparison of the leaching data of veterinary pharmaceuticals obtained according to different procedures.

Generally, laboratory leaching tests can be divided into static and dynamic experiments. Static tests are commonly used when the objective of the study is to calculate the distribution coefficient K_d [L kg^{-1}], in equilibrium time, which is specific to every chemical. Dynamic tests e.g. column tests, on the other hand, enable time-dependent monitoring of contaminant leaching from soil and waste materials; in addition, the flow-through pattern of such tests resembles actual environmental conditions. Although both column and batch tests can be used to assess the possible leaching/release potential of contaminated materials on the soil–groundwater pathway (Delay et al. 2007; Grathwohl and Susset 2009; Grathwohl and van der Sloot 2007), a batch test only provides a snapshot of a particular liquid-to-solid ratio, whereas during a column leaching test the release of a contaminant depends on the local equilibrium time and advection conditions (Lopez Meza et al. 2008). Such an assessment provides more environmentally relevant results eliminating the possible overestimation of sorption present in batch tests. Therefore, the dynamic test seems to be more appropriate in the groundwater contamination risk assessment of VPs.

3.1 Literature Data on VP Leaching

The trend of shifting from batch leaching tests towards column experiments was noticed more than a decade ago. Grathwohl and Susset reported that batch tests are not a good tool to mimic environmental scenarios of leaching behavior due to under-or over-predicted concentrations in water phases (Grathwohl and Susset 2009). However, current understanding of the leaching potential of veterinary antibiotics into groundwater is still incomplete. Due to its simple experimental setup, the batch test has frequently been used and equilibrium sorption coefficients for various VPs have so far been determined.

For the following groups of VP_S, the sorption coefficients are well known: TC_S, SA_S, AG_S, MQ_S, FQ_S, IMD_S, polypeptides, polyethers, glycopeptides and QXL_S (Song and Guo 2014). Although their distribution coefficients (K_d) depend strongly on a number of factors (e.g. soil parameters like organic matter content, physico-chemical properties of the chemical compound, and external factors such as pH, ionic strength or temperature) on the basis of the presented ranges a strong interaction with soil particles is noticeable for TC_S and FQ_S (420–1030 and 260–6310 L kg^{-1}, respectively) whereas relatively weak interactions are exhibited by e.g. SA_S [0.6–107.5 L kg^{-1} (Białk-Bielińska et al. 2012)] and IMD_S (0.54–0.67 L kg^{-1}) (Song and Guo 2014). Therefore, SA_S and IMD_S are likely to migrate out of a field in runoff water or can be leached down through the soil profile by percolation water. On the other hand, TC_S are considered strongly sorbed by soil particles with K_d, reported to be as much as >300,000 L kg^{-1} for clay soil-and thus are frequently treated as scarcely mobile in this compartment of the environment (Sassman and Lee 2005; Schlabach et al. 2007). Several studies have reported small amounts of pharmaceuticals or even their lack in leachates, whereas OTC and TC have been detected in soil (Popova et al. 2014; Kay et al. 2005a; Kim et al. 2010). However, strongly sorbed chemicals can also move to other locations together with associated soil particles when eroded by runoff water. Indeed, Davis et al. reported very low aqueous concentrations of TC and CTC during a runoff study; however, their relative losses including particle facilitated transport were 65 and 34%, respectively (Davis et al. 2006). This means that sorption in this regard does not make chemicals immobile in the environment.

Aust et al. observed particle-facilitated transport and macropore flow of CTC resulting in translocation from manure layers into mineral soil. They also indicated that horizontal transport of SMT and CTC is possible due to surface runoff. Moreover, both horizontal and vertical flow within feedlots may cause the dislocation of antibiotic residues into both surface and groundwater resources (Aust et al. 2008). Another study reported that, although veterinary antibiotic residues (*i.a.* TC_S and their degradation products) mainly stayed on the surface after the spreading of swine manure and a subsequent rainfall, these residues were to some extent also distributed in the various soil depth layers of the field, signifying possible translocation in the nearby environment via drainage waters (Solliec et al. 2016).

In their recent study, Sura et al. indicate bulk density as a crucial factor affecting the surface runoff of veterinary antimicrobials. They examined the transport of CTC, SMT and TYL, which are commonly used in beef cattle production, in simulated rainfall runoff from feedlot pens and observed that increased bulk density lowered infiltration and finally caused a higher runoff volume (Sura et al. 2015). Surface water contamination results from surface runoff or drain-flow; hence, it is possible to mimic this process in the field, as described above. However, it is much more complicated to observe leaching to groundwater via infiltration through soil profiles under field conditions. Although such scenarios can be simulated in the laboratory with batch and column tests, there are discrepancies in the results obtained according to different methodological approaches. For example, the results obtained in batch and column tests for NA showed a lack of agreement (Clervil et al. 2013). Although the order of affinity determined by the batch results was consistent with that observed in the column system, the retardation factor derived from batch data was higher than that determined from the column breakthrough curves for all the sediments tested. Moreover, a discrepancy in terms of the total sorbed amount was also observed. The authors highlighted that this could be attributed to several factors linked to the differences between the experimental conditions of batch and column investigations: solid/liquid ratio, the loss of sorbent particles through transport, variations in column flow and moisture content, column flow channeling and the presence of immobile water regions as well as the kinetic behavior of adsorption (Clervil et al. 2013). Liu et al. also reported that, in the case of weakly sorbed compounds, the failure to incorporate sorption kinetics into pharmaceutical transport during water quality modeling leads to a significant underestimation (80%) of dissolved-phase concentrations, whereas for compounds exhibiting fast sorption, an overestimation of up to 150% may occur (Liu et al. 2013b). Although the leaching of SA_S was comparable in batch and column tests in the study performed by Maszkowska et al., the authors observed slight differences in leaching determined within batch testing in comparison to column testing (Maszkowska et al. 2013). Greater leaching was observed for the column test in the case of soil with a considerably higher clay fraction content. This has been explained by the different test procedures and the possibility to produce more fine particles during tumbling in the batch test and thus enhanced sorption. For coarse-grained soil, greater leaching was observed in the case of the batch test. This could have arisen from the low probability of producing more very fine particles and hence no possibility of reduced leaching via colloid sorption. A correlation between hydrophobicity and leaching potential was also observed. The slowest release was reported for SDM—the most hydrophobic among the tested compounds (24.5% decrease in concentration to L/S 1.22 L kg^{-1}).

A comparison of batch and dynamic tests was also performed by Kurwadkar et al. A column test showed that antimicrobials such as the selected SA_S, TYL and CBX achieved only a nominal 40–50% of equilibrium capacity. The results in the column tests therefore reflect transport under conditions of partial equilibrium, as may occur under typical storm conditions in the field (Kurwadkar et al. 2007, 2011). Among the investigated compounds, SA_S were more mobile than TYL and CBX, which in contrast were considered immobile with a low risk of reaching the groundwater. However, relatively mobile SA_S also turned out to be immobile in slightly acidic soils with greater organic carbon and a greater cation-exchange capacity (Kurwadkar et al. 2011).

As already mentioned above, TC_S exhibit the potential to be translocated via macropore flow, though their K_d coefficients are relatively high. Ostermann et al. also confirmed this conclusion (Ostermann et al. 2013). Although only TC and DC among the investigated TC_S occurred in the leachate at a single event, an initial preferential flow was reported to be probable for these compounds under field conditions. Preferential flow comprises the physical phenomena of the rapid transport of water and solutes in soil bypassing other volume fractions of the porous soil matrix and is attributed to flow through cracks, fissures, and biophores, e.g. earthworm burrows and root channels. Additionally, preferential flow can arise from non-homogenous infiltration or wetting flow instabilities (Bundt et al. 2001; Gerke 2006). Thus, even highly sorbed compounds such as TC_S may be more mobile than anticipated and pose a contamination risk to surface or ground waters. Nevertheless, the occurrence of TC and DC in the leachate was considerably lower than the concentration of SA_S (Ostermann et al. 2013).

Blackwell et al. showed that OTC is unlikely to move to groundwater, but has some potential to move to surface water within a short period after application. Moreover, the concentrations of OTC in runoff were at sub μg L^{-1} levels and these would drop significantly when the runoff was diluted in surface water bodies (Blackwell et al. 2007). The conclusion of Popova et al. about the relatively low risk of the significant transport of CTC, OTX, and IVM to surface-and ground waters from cattle manure on irrigated pasture tallies with this (Popova et al. 2014). The leaching of SA_S via preferential flow paths is in agreement with previous data for sulfachloropyridazine (SCP) which indicate a preferential flow for this SA in both undisturbed clay loam, where macropores are likely to exist (Kay et al. 2005b), and homogeneous sandy loam (Blackwell et al. 2009). SCP showed the potential to move to both groundwater and surface water in the study of Blackwell et al. (2007), but at low or sub μg L^{-1} concentrations before dilution into the receiving water (Blackwell et al. 2007). Ostermann et al. also determined only small portions of antibiotics, mainly SA_S, in the leachates. However, taking into consideration the fact that their concentrations in leachates constantly exceeded the threshold-value for groundwater contamination in the worst-case scenario, they indeed pose a risk for gradual groundwater contamination in the long run (Ostermann et al. 2013).

Despite all the advantages of column leaching tests, especially more environmentally relevant experimental conditions ensuring leaching data is not underestimated due to overestimated sorption results, their weakness also needs to be pointed out. For example, Kay et al. emphasized several factors leading to great discrepancies between a lysimeter study and a field study, such as a low representation of the spatial heterogeneity of the field soil in lysimeters in terms of hydrology, a lack of backfill in the lysimeter, and a different temperature (Kay et al. 2005b). Moreover, the influence of temperature on the sorption potential of ionizable pharmaceuticals onto soil particles has already been proved in the literature (i.a. SA_S) (Maszkowska et al. 2014). Although lysimeters may not be subject to all of the processes occurring in the field, according to Kay et al. this problem may be solved by collecting relatively large lysimeters and/or incorporating important hydrological pathways, such as drain backfill and desiccation cracks, into the lysimeters (Kay et al. 2005b). Also, test material has a strong influence on the obtained results. According to Oppel et al., topsoil and subsoil may differ significantly in terms of organic carbon, bacterial community, and further properties. Therefore, the results from leaching tests performed with topsoil have only a limited transferability for subsoil where there may be a source of groundwater contamination (Oppel et al. 2004). Moreover, in natural soil environments, when unsaturated flow conditions as well draught periods may lead to the drying out of the soil, leaching can be decreased by moving the chemicals back towards the soil surface as a result of capillary forces (Rabølle and Spliid 2000).

3.2 *Occurrence of VPs in Ground Waters and Comparison to Leaching Data*

The fast leaching of SA_S in laboratory tests or field studies is supported by the results of groundwater monitoring studies. A range of these antibiotics, e.g. STD, SMT, SDM and SMX, have been detected in different studies in groundwaters in a concentration of up to 3461 ng L^{-1} (Barnes et al. 2008; Bartelt-Hunt et al. 2011; Diaz-Cruz et al. 2008; Estévez et al. 2012; García-Galán et al. 2010; Lapworth et al. 2012; Lindsey et al. 2001; Sui et al. 2015; Teijon et al. 2010). Moreover, García-Galán et al. detected 18 out of the 19 target SA_S in two groundwater bodies from Catalonia. The highest frequency of detection was observed for SMT and SDM (89.4%); however, an acetylated metabolite of SMT also occurred frequently in the collected samples (82%) (García-Galán et al. 2010). These SA antimicrobials have also been detected by Batt et al. in groundwater samples of private water wells in USA. The concentration ranges were 76–215 and 50–68 ng L^{-1} for SMT and SDM, respectively. Thus, the negative impact of a concentrated animal feeding operation (CAFO) on the quality of the local groundwater has been proved (Batt et al. 2006).

While the presence of SA_S in water compartments, including groundwater, is justified, the leaching of TC_S is generally considered in the literature as being unlikely to occur. However, López-Serna et al. reported the occurrence of TC and OTC in groundwater underlying the city of Barcelona in Spain. Moreover, the concentrations determined were comparable to those obtained for SA_S (López-Serna et al. 2013). In the same study, the fluoroquinolone antibiotic—NOR, occurred in groundwater in the relatively high concentration of 462 ng L^{-1}, although its sorption coefficient indicates low mobility in soil (Conkle et al. 2010). However, mobility studies are normally performed in laboratory conditions while in a real environmental situation the sorption can be disturbed by the other matrix components obtained in the soil and transmitted water. Other VP_S, the presence of which in groundwaters has been determined, are as follows: TMP, ROX, SPM, MTZ, AZM, ERY, LIN, and MON (Barnes et al. 2008; Bartelt-Hunt et al. 2011; K'oreje et al. 2016; López-Serna et al. 2013; Sui et al. 2015).

As has already been mentioned above, the potential transport of veterinary medicines into groundwater is recognized by EU guidance documents. Therefore, in order to ensure data for a large set of chemicals and various scenarios, different models for the estimation of the predicted exposure concentrations (PEC) in relevant environmental compartments (e.g. groundwater) are available. The VetCalc model and the Forum for the co-ordination of pesticide fate models and their use (FOCUS) such as PEARL (Pesticide Emission Assessment at Regional and Local scales), PELMO and MACRO are recommended for such assessments of VP_S (Metcalfe et al. 2008). Concentrations in groundwater are approximated by pore water concentrations (i.e. $PEC_{groundwater} = PEC_{porewater}$), and they can be derived according to equations provided in the guidelines on the basis of predicted environmental concentrations in soil (PEC_{soil}). Although such models often use real experimental data instead of predicted data (e.g. sorption coefficients on the basis of octanol-water partition coefficients in PEC_{soil} calculations), discrepancies and gaps concerning the fate of VPs sometimes cause these results to diverge from reality.

Blackwell et al. evaluated the predictive capability of the PEARL leaching model. A comparison of the experimental measurements with simulations from this model demonstrated that it greatly underestimates the transport of antibiotics to groundwater. SCP, which is recognized as a mobile compound, and the occurrence of which in groundwater has been reported in many experimental studies, is considered by the model to be not leachable (Blackwell et al. 2009). Thus, leaching models can strongly underestimate the leaching behavior of VPs, probably due to their inability to consider factors affecting leaching in the field [e.g. different degradation rates due to temperature changes or differences in bioavailability (Blackwell et al. 2009)]. On the other hand, the study of Boxall et al. indicates the greatly overestimated risk of pharmaceuticals the occurrence of which in groundwater has already been reported at much lower concentration levels. The $PEC_{groundwater}$ for MON (326 μg L^{-1}) investigated in this study was revealed to be overestimated around 800-fold in comparison to concentrations seen in the field (0.39 μg L^{-1}) (Boxall et al. 2006b). Moreover, even an additional study determining VPs in groundwater impacted by a large CAFO detected MON in much lower concentrations (2.35 μg L^{-1}) than was the case for $PEC_{groundwater}$ (Bartelt-Hunt et al. 2011).

An overestimated PEC_{soil} has also been highlighted by Metcalfe et al. If concentrations of veterinary medicines in manure were either measured or spiked, there was better agreement between predicted and measured concentrations in soil (Metcalfe et al. 2008). A possible explanation for this lay in the overestimated animal husbandry scenarios and manure application scenarios in comparison to actual inputs to soils. In contrast, Blackwell et al. showed that the PEARL model seemed to work well in identifying compounds that do not leach (i.e. TC_S). However, further investigation of groundwaters demonstrated OTC occurrence in groundwater at a concentration of 41 ng L^{-1} (López-Serna et al. 2013). A possible reason for this could be attributed to the inappropriately low limit of detection/ quantification of analytical methods used so far for groundwater monitoring. Sensitivity is, therefore, one of the most critical parameters in order to obtain unequivocal and reliable determinations for groundwater risk assessment. Moreover, due to analytical constraints, the concentrations used in sorption studies are often significantly higher than those that would normally be encountered in the environment. Under field conditions, where concentrations of VPs are expected to be at the ng L^{-1} level, greater sorption capacities may be exhibited due to e.g. binding to specific and rare high-energy adsorption sites, which cannot be observed under laboratory conditions with relatively high concentrations of sorbate (Borgman and Chefetz 2013; Srinivasan et al. 2014). Therefore, even if in leaching modeling the sorption coefficient determined in batch testing (not the predicted one) is used, the leaching modeling could under or over–estimate the leaching behavior of VP_S. Srinivasan et al. proposed the concentration-dependent effective distribution coefficient (K_d^{eff}) as the basic tool for leaching estimations, as it provides a more realistic field situation (Srinivasan et al. 2014).

Another reason for underestimated modeling results for strongly sorbing compounds could be particle-facilitated transport, which is not considered in the simple algorithms suggested by the Committee for Veterinary Medicinal Products (Metcalfe et al. 2008). Possible discrepancies between modeling and field studies may also arise from differences in the scenario that is modeled and the systems where the monitoring has been done. Taking all the above mentioned into account, there is still work to be done to improve existing models as well as the analysis which is necessary for the sensitive detection of VPs at trace levels in complex matrices. Moreover, laboratory leaching experiments still need to be conducted in order to provide large sets of experimental data, especially regarding factors influencing the sorption and mobility of VPs in soil. It is also worth stressing at this point that it seems invaluable to combine data from models and monitoring studies including climate factors and regional scenarios. One such comprehensive approach recently proposed for groundwater risk assessment from pesticides (Di Guardo and Finizio 2016) would also greatly improve risk assessment related to the leaching of VP_S.

4 VPs vs. Soil Microorganisms

Among all the groups of VPs, antibiotics exert significant influence on soil microorganisms. As they are designed to selectively affect microorganisms, once introduced to soil they might then affect the structure and function of bacterial communities as well as the development and spread of antibiotic resistance. TC_S (including CTC, OTC, and TC) and SA_S (including SMT, SMX and SDZ) are typical antibiotics that are widely used in animal husbandry; thus, most literature data on the impact of veterinary antibiotics on bacteria in soil environments concern SA_S and TC_S.

4.1 Toxicity Toward Soil Bacteria and Fungi

The available data on the toxic effect of VPs on the environment and, in particular, soil bacteria are scarce. It has been documented that, given the nature of the soil environment, it is impossible to evaluate toxicity toward soil bacteria based on results obtained in tests for single species (such as growth inhibition tests with *Pseudomonas putida* or bioluminescence inhibition tests with *Vibrio fischeri*) and environments other than soil. For example, the inhibitory effect of OTC on activated sludge bacteria occurs at significantly lower concentrations (EC50 = 0.14 mg L^{-1}) (Halling-Sørensen 2001) compared with soil bacteria (EC50 = 50 mg kg^{-1} dry soil) (Vaclavik et al. 2004). The growth inhibition of soil bacterial isolates has been observed in agar diffusion disk tests for CTC and SDZ, whereas in microcosms with soil spiked with these antibiotics no effects have been observed up to the highest tested concentrations of 50 mg kg^{-1}, either on respiratory activity or on bacterial population structure (Zielezny et al. 2006).

Numerous studies have documented changes of in soil microbial community structure as a result of exposure to antibiotics in the environment. Most often, molecular fingerprinting techniques reveal changes in band patterns or in intensities of signals representing different microbial species (Hammesfahr et al. 2008; Reichel et al. 2013; Zielezny et al. 2006). The long environmental residence times of FQ_S in soil matrices affect the composition and activity of soil microbial communities, which was demonstrated for CIP at 200 μg kg^{-1} by Girardi et al. (2011). Schmitt at al. reported that the contamination of soil with SCP resulted in small changes in the community level physiological profile (CLPP) accompanied by increased pollution induced community tolerance (PICT) against further additions of the sulfonamide. A decrease in the microbial diversity indexes at 100 μg L $^{-1}$ of TC was observed in a batch reactor experiment (Zhang et al. 2013a). The microbial community structure of wheat rhizosphere soil exposed to different concentrations of OTC was highly affected. It was shown that exposure to OTC inhibited the growth of *Bacillus* and therefore results in a decline in soil enzyme activity in the wheat rhizosphere (Yang et al. 2009). In contrast, Berglund et al. observed no effect on bacterial diversity after continuous exposure to a mixture of 12 antibiotics in experimental wetlands (Berglund et al. 2014). Similarly, no significant differences were observed in constructed wetlands in terms of bacterial abundance, richness or diversity, among different treatments (without a pharmaceutical addition or with TC or with ENR at concentrations of 100 μg L^{-1}) or throughout the duration of the experiment. However, a significant effect of both time and treatments was noted on bacterial community structure by Fernandes et al. (2015). A high level of diversity is considered to be an important feature of ecosystem integrity, as it implies functional redundancy, acting as a genetic and functional reservoir that increases community resilience to disturbance (Bissett et al. 2007). Therefore, a loss of community diversity has been used to indicate a decline in ecosystem function (Allison and Martiny 2008). It has been shown that changes in microbial community structure can affect the ecological functions of soil ecosystems, e.g. biomass production and N-transformation processes (Kotzerke et al. 2008). Nevertheless, other studies have revealed that community shift is not necessarily mirrored by altered soil functioning, but is masked by functional redundancy sustained by a structurally changed microbial community (Fernandes et al. 2015; Hammesfahr et al. 2008).

Additionally, the ratios of bacteria/fungi and Gram positive/Gram negative bacteria, the two indicators most often used to describe changes in microbial community structure, were studied. An increased fungi/bacteria ratio was reported after the application of manure containing SDZ (at a concentration of 10–100 mg kg^{-1}) in soil, due to both a reduction in bacterial biomass and a stimulated growth of fungi (Hammesfahr et al. 2008). Other antibacterials, such as SPY, OTC (Thiele-Bruhn and Beck 2005) and SMX (Demoling et al. 2009), also increased fungi/bacteria ratios. A shift towards Gram positive bacteria and fungi was observed after the application of slurry from SDZ and DFX medicated pigs (Reichel et al. 2013). It is noteworthy that antimicrobial agents may also act on non-target organisms; OTC and PEN, when combined together, reduced the lengths of active hyphae of fungi in forest soils (Colinas et al. 1994).

Bacteria form one of the most important groups of organisms in soil and are essential for the decomposition of organic matter, and geochemical cycles such as the carbon, nitrogen, sulfur and phosphorus cycles. Knowledge of the effects of antibiotics on the ecological functions of soil is still limited. Three reports have described the influence of SDZ on soil nitrification/denitrification processes (Kotzerke et al. 2008; Ollivier et al. 2010; Schauss et al. 2009). SDZ affects the abundance and diversity of nitrifying bacteria and leads to a decrease in ammonia-oxidizing bacteria and an increase in ammonia-oxidizing archaea (Ollivier et al. 2010; Schauss et al. 2009). Apart from the nitrification process, Kotzerke et al. also studied the effects of SDZ on denitrification processes in soil. With SDZ at 10–-100 mg kg^{-1}, the denitrification rate in the soil samples was significantly reduced (Kotzerke et al. 2008).

The antibiotic effects on microbial activities and enzymes in soil have also been reported. The influence of various antibiotics, such as CIP, SDZ (Kotzerke et al. 2008; Zielezny et al. 2006), TYL (Demoling and Bååth 2008; Muller et al. 2002), CTC and SCP (Vaclavik et al. 2004) on the soil respiration rate and soil biomass production has been reported. OT effects on soil microbial respiration have been determined in soil with or without the addition of manure. In all cases, an NOEC of 10 mg kg^{-1} was obtained (Boleas et al. 2005). In contrast, Hund-Rinke et al. (2004) did not find any influence of TC on soil respiratory activity with concentrations up to 50 mg kg^{-1} soil. The inhibition of alkaline phosphatase after exposure to 10–-30 mg kg^{-1} OTC in soil was observed, while the other three soil enzymes tested (acidic phosphatase, dehydrogenase, and urease) were not affected (Yang et al. 2009). Thiele-Bruhn and Beck reported that dehydrogenase activity and basal respiration in soil were not significantly affected by either SPY or OTC, regardless of the doses of antibiotics. The lack of any observed effect could be, according to the authors, explained by shifts in the microbial community structure that compensated for effects on single species. In other studies, Thiele-Bruhn tested nine pharmaceutical antibiotics for their effects on the reduction of microbial iron(III) in six different topsoils. The derived effective doses (ED [μmol kg^{-1} soil]) for the different antibiotics increased in the following order: CTC (53) <SDM (58) < OTC (170) < SDZ (190) < SMD (270) = TC (270) < SPY (430), although no effect was found for SFL and FBZ at doses of up to 5.8 and 3.3 mmol kg^{-1}, respectively. Both the inherent chemical properties of antibiotics and environmental parameters could influence the actual effects of antibiotics on the soil microbial community. The effects of antibiotics are influenced by their actual concentrations in soil that could come into direct contact with microbes (bioavailable concentration). The actual concentration in the soil and aqueous phase is a function of time, stability or the persistence of the antibiotic, soil properties (such as soil pH, soil organic matter, and soil minerals), and microbial activity (Thiele-Bruhn 2003). A number of studies have shown that sorption may lead to the reduction (Hund-Rinke et al. 2004; Kotzerke et al. 2008; Thiele-Bruhn 2005) of antibiotic effects on microbial communities. Determined effective concentrations (EC50) in soil solution derived from sorption experiments are considerably smaller and range from 0.004 μmol L^{-1} (CTC) to 120 μmol L^{-1} (SPY) (Thiele-Bruhn 2005).

The consequences of the observed changes in community structure influence the function of the soil and it is expected that such disturbances might have significant and long-term effects on ecosystem homeostasis. However, long-term studies frequently detect a recovery of the community biomass and a growth of certain microbes that were initially inhibited. Demoling et al. (2009) employed PLFA and pollution-induced community tolerance (PICT) analysis to evaluate the effects of SMX on soil microbes, which demonstrated an initial decrease in bacterial growth rates and a gradual amplification of more tolerant species, as reflected by an increase in PICT. Similarly, in Kotzerke et al. (2008), microbial activity recovered at day 32 when a high concentration of SDZ was added.

4.2 Resistance of Bacteria

The application on land of manure from antibiotic-treated animals has been reported to frequently increase not only the levels of antibiotics, but also the abundance of resistant bacteria and antibiotic resistance genes (ARGs) in soils (Ghosh and LaPara 2007; Martínez 2008; Białk-Bielińska et al. 2014). Resistance genes, as well as resistant bacteria in the environment are increasingly seen as an ecological problem and, furthermore, are considered to be new contaminants which might pose a potential worldwide human health risk (Zhu et al. 2013). The diversity and abundance of various ARGs has been detected in soil with the application of swine manure in different countries (Heuer et al. 2011; Huang et al. 2013; Wu et al. 2010a, b).

The development of resistance is believed to be promoted by continuing exposure to sublethal concentrations of antibiotics (Thiele-Bruhn 2003). SA_S concentrations as low as 0.1 mg kg^{-1} of soil could have a selective effect on resistant populations in soil (Heuer et al. 2008). The extent of the increase in resistance in soil following manure fertilization is related to the intensity of manure fertilization (Karci and Balcioğlu 2009; Sengeløv et al. 2003). A study of agricultural soil from The Netherlands has shown the accumulation of antibiotic resistance genes over six decades of increasing use of antibiotics (Knapp et al. 2010). Nine classes of *tet* genes are dominant in arable soil after 6 years of the application of fresh or composted manure (Peng et al. 2015). Fang et al. (2014) reported that bacterial community tolerance to CTC and SDZ in manure-amended soil increased significantly with the frequency of antibiotic treatment. Following three sequential applications of manure containing antibiotics, the accumulation of numbers of copies of sulI and sulII significantly increased compared to the treatment with antibiotic-free manure or unfertilized control soil (Heuer et al. 2011). Changes in community tolerance after exposure to a certain pollutant can be measured by PICT analysis. PICT changes under antibiotic treatment not only reflect shifts in microbial community structure, but also imply the amplification of antibiotic resistance within a community (Ding and He 2010). Several studies found substantial PICT increases after soil was treated with SA_S, e.g., SDZ (Brandt et al. 2009), SMX (Demoling et al. 2009) or SCP (Schmitt et al. 2005) and TYL (Demoling and Bååth 2008).

The resistance of microorganisms is not only provoked by the input of antibiotics into the environment and their selective pressure. It appears to be more important that resistant microorganisms co-occur with manure introduced into soils. However, it was demonstrated that manure-derived bacteria could not thrive in soil environments, and gradually decreased after manure treatment (Hammesfahr et al. 2008; Heuer et al. 2008), which was attributed to differences in environmental conditions between the soil and animal gut. It is known that the survival times of culturable fecal bacteria in soil are only in the range of weeks to months (Chee-Sanford et al. 2009). Since bacteria from manure are not sufficiently adapted to soil environments, horizontal gene transfer from manure to indigenous soil bacteria might be important for the dissemination of resistance in soil (Chee-Sanford et al. 2009; Gillings and Stokes 2012; Götz and Smalla 1997; Heuer et al. 2011). Most ARG cassettes are found on integrons frequently located on plasmids and transposons, which might be transferred from manure bacteria to soil bacteria (Binh et al. 2008; Allen et al. 2010; Heuer et al. 2011). Plasmids belonging to the groups Inc-P1, IncQ, IncW, and IncN, which play an important role in disseminating antibiotic resistance genes, have been identified in both manure and farm soil samples (Binh et al. 2008; Heuer et al. 2011; Smalla et al. 2000). Also, an increased activity of integrases and transposases, enzymes related to transposition processes, has been observed. Resistance genes located on broad host range plasmids make transfer between distantly related species more possible (Heuer et al. 2011; Zhu et al. 2013). Additionally, LowGC-type plasmids have been observed to be highly abundant in manure and manured soil. Some antibiotic resistant bacteria in soil and manure are phylogenetically close to human pathogens (for example, Acinetobacter spp.-identified as a potential host for LowGC plasmids), making genetic exchange more likely (Byrne-Bailey et al. 2009). Moreover, it has been shown that broad host range resistance plasmids belonging to the IncP-1ε group are frequently captured from manure and manure treated soil into E. coli recipients (Heuer and Smalla 2012).

Microbes containing the ARGs from manure or soil are possibly subject to dispersal via leaching to subsurface soils and groundwaters (Pruden et al. 2012). Sulfite-reducing TC resistant clostridia have been detected in manure, manure fertilized soils and groundwater (Huysman et al. 1993). It was reported that, in contrast to the decreasing concentration of antibiotics, four ribosomal protection protein genes (tetM tetO tetQ tetW) increased with soil depths and the gene copies normalized to 16S rRNA even reached 10^{-4} or 10^{-3} in the deepest 40–60 cm layer (Tang et al. 2015). The emergence and spread of ARGs in different depths of soil may also be closely associated with mobile genetic elements (Binh et al. 2008; Heuer and Smalla 2012; Zhang et al. 2011). Huang et al. (2013) reported that the resistance determinants can migrate to deeper soil layers and could possibly contaminate the groundwater by vertical transport. On the other hand, it has been documented in manure amended plots that the abundance of ARGs in surface soils is orders of magnitude higher than in subsoils (Joy et al. 2014). Moreover, Fahrenfeld et al. (2014) reported that no evidence existed for the surface or subsurface transport of ARGs in soil due to manure application using a field-scale mass balance approach.

There is little information on how quickly a reduction in antibiotic use will result in decreased resistance. According to Tamminen et al. (2011), once established by the selective pressure of antibiotics, resistance genes persist even after the selective pressure is removed. The prevalence of SA resistance genes (*sulI* and *sulII*) was observed in bacterial isolates from agricultural soils in the United Kingdom after 2 years of the application of manure containing antibiotics, compared to non-amended soil (Byrne-Bailey et al. 2009). Hong et al. (2013) reported an increased abundance of *tet* genes in soil after pig manure application, and these levels remained elevated for up to 16 months. A reduction in resistance levels was reported in the literature; however, it was only in the range of percentages, and resistance genes persisted for many years in the absence of the corresponding antibiotic compounds (Johnsen et al. 2009). Acquired antibiotic resistance genes and mobile genetic elements often impose a cost on the host cell, which reduces that host's fitness in the absence of antibiotics. It is also possible that the cost of acquired resistance genes in the absence of selective pressure is further reduced at the population level by heterogeneous permissiveness for horizontal gene transfer. Resistance genes persist in a permissive subpopulation that increases in relative abundance in cases of antibiotic selection (Heuer et al. 2011).

The survival of microorganisms in the presence of antibiotics is not only due to acquired resistance. Soils represent habitats and sources of indigenous antibiotic-producing microorganisms (Thiele-Bruhn 2003). Natural antibiotics produced by bacteria and fungi occur in the environment and control the dynamics of microbial populations. Such antibiotics synthesized in situ are found especially in the soil rhizosphere, with concentrations of up to 5 μg g^{-1} (Thiele-Bruhn 2003). Among numerous other soil organisms, 30–50% of actinomycetes isolated from soil are able to synthesize antibiotics (Allen et al. 2010). Antibiotic producers contain resistance genes within a biosynthetic gene cluster to avoid autoinhibition (Cundliffe 1989). Many investigations have revealed that numerous soil microorganisms have a natural tolerance towards antibiotics. Out of 36 strains of microorganisms from uncontaminated soil and water, only seven were susceptible to 21 diverse antibiotics (Thiele-Bruhn 2003). In particular, pseudomonas are often intrinsically resistant to antibiotics (Halling-Sørensen et al. 2003a, b). Therefore, it is not surprising to detect a set of diverse ARGs in both untreated soils and control pristine soils; and these ARGs represent nearly all the classical resistance mechanisms including antibiotic efflux, target protection, and antibiotic inactivation (Walsh 2000). Moreover, it has been shown that resident soil antibiotic resistance bacteria are also found to significantly increase following the application of manure from cows without a history of antibiotic treatment (Udikovic-Kolic et al. 2014). The substrates from manure may permit the growth of microbial populations carrying *tet* and *sul* genes (Heuer et al. 2011). The addition of nitrogen fertilizer may strongly influence the content of ARGs in soil (Forsberg et al. 2012). Other factors that might indirectly enhance the spread of antibiotic resistance are metals which accumulate in soil owing to manure application. Cu, for instance, has been shown to co-select for antibiotic resistance in microbial soil communities under field conditions (Berg et al. 2010). However, it has been shown that resistance in pristine soils differs quantitatively and qualitatively from soils under agricultural influence. For example, in soil samples from the Rocky Mountain National Park the TC resistance genes tetB, tetC, tetW, and tetO are not detected by real-time PCR, while these genes are abundant in soil samples from other sites in Colorado affected by agricultural or urban activities (Jiaa et al. 2014).

Soil is considered to be the largest environmental reservoir, comprising as much as 30% of the known ARGs in public repositories (Nesme et al. 2014). High-throughput functional metagenomic analysis has found that soil bacteria harbor resistance gene cassettes against all major classes of antibiotics with high levels of similarity to genes from human pathogens (Forsberg et al. 2012). As the only ecosystem interacting constantly with all compartments of the biosphere, soil is prone to genetic exchange by means of horizontal gene transfer between ecologically distinct lineages found in other ecosystems (Nesme and Simonet 2015). The pollution of the environment by veterinary antibiotics and antibiotic resistant determinants likely increases the chance of human-associated bacteria and human pathogens acquiring resistances by mobile genetic elements such as broad host range plasmids, facilitating horizontal gene transfer between distantly related species (Heuer et al. 2011).

5 VPs vs. Plants

Manure, both solid and liquid, is normally distributed onto residential allotment gardens (small scale), as well as crop plantations (large scale). This manure can be composted before use to increase the bioavailability of nutrients, improve the physical and chemical properties of composts and suppress phytopathogens (Moral et al. 2009). A side effect is the degradation of pharmaceutical bulk [more than 99% of the antibiotic removal from manure during 40 days of composting (Ho et al. 2013)].

VPs can be found in water (ng L^{-1}), soils (level of ng kg^{-1} to μg kg^{-1}) and manure/biosolids (ng kg^{-1} to mg kg^{-1}) (Ding et al. 2011; Dorival-García et al. 2015; Ho et al. 2012; Kemper 2008; Solliec et al. 2016; Wei et al. 2016), and all of these can affect the crop plant. Studies focusing only on the impact of VP_S on plants are scarce. Most studies have investigated the intake of pharmaceuticals from reclaimed wastewater or sewage sludge as a fertilizer (Carter et al. 2015; Cortés et al. 2013; Fatta-Kassinos et al. 2011; Grassi et al. 2013; Miller et al. 2015; Shenker et al. 2011; Verlicchi and Zambello 2015; Wu et al. 2013). The plants which fall within the scope of this study are edible crops, e.g. carrot, radish, lettuce, cucumber and tomato, onion (Carter et al. 2014; Chuang et al. 2015; Herklotz et al. 2010; Holling et al. 2012; Kang et al. 2013; Kumar et al. 2005; Sabourin et al. 2012), and also soya and cereals (Hawker et al. 2013; Marsoni et al. 2014). Most of these can be consumed without the need for processing.

The interactions between plants and pharmaceuticals have been under investigation for several years, because of the use of phytoremediation in constructed wetlands (Carvalho et al. 2014). Phytoremediation is a technology that utilizes plants and the associated rhizosphere microorganisms to remove, transform or contain toxic chemicals located in soils, sediments, groundwater, surface water and even the atmosphere (Susarla et al. 2002). Several of the studied plants have been shown to be effective in the phytoremediation of veterinary and human pharmaceuticals (Carvalho et al. 2014). This process is not only connected with plant uptake, biodegradation and photodegradation, but it is also important in pharmaceutical elimination (Zhang et al. 2014a, b). The ratio of these three elimination methods varies between pharmaceuticals (Matamoros et al. 2012; Zhang et al. 2013a, b), but the exact values for VPs are as yet unknown.

5.1 Uptake and Translocation

Generally, water and small solutes (Mr $\leq$ 500) 108 can enter the root through the epidermis (Trapp and Mc Farlane 1995); they can then be transported via the xylem/phloem to internal tissues, provided they are not stopped by barriers inside roots (Miller et al. 2015). The chemical character of neutral compounds, mainly the lipophilicity, positively affects the ability to cross plant cell membranes by passive processes (Sterling 1994). The surface potential of the cytoplasmic membrane is negative, thereby charged molecules will interact with it (Devillers 2009). This leads to the accumulation of compounds within plant root cells. Some pharmaceuticals are expected to be subject to ion trapping, but experimental studies have not been carried out (Miller et al. 2015). Ion exchange of negatively charged walls and cationic pharmaceuticals can decrease their pathway to underlying cells, but currently there is no information to prove this statement. Protein-mediated transport, observed in the case of some plants, is another possible means of pharmaceutical uptake. Many organic nitrogen transporters obtained in plants have low selectivity (Rentsch et al. 2007), which suggests they can mediate in the transport of those VPs which are neutral and structurally similar to the originally transported compounds. Quaternary ammonium compounds can be taken up by plants as well (Warren 2013), but actually no VPs have a quaternary ammonium character. Some of the assimilated molecules rich in nitrogen can have high masses, of several thousand Da, so uptake is not limited to the low mass-compounds. Given that some molecules cannot enter the deeper layers they will stop in the roots. This has been proven for trimethoprim and sulfonamides (Tanoue et al. 2012). Accumulation in roots is in contrast to the translocation of VPs in plants to leaves, seeds and fruits. More information can be found in the review by Miller et al. (2015).

Since some plant roots represent a destination place for many pollutants, their accumulation in plants can be expressed using the root concentration factor (RCF), i.e. the ratio of the concentration in roots to the concentration in the exposure medium (soil or soil pore water). For example, the RCF of OTC and NCR in soybean roots are 0.68–3.32 and 0.16–1.52 (Boonsaner and Hawker 2010). The uptake factor (UF, concentration of an analyte in plant material to its concentration in soil) is more general than the RCF and more reliable than the BCF (concentration of an analyte in plant material to its concentration in water used for irrigation). The UF for lettuce and carrot is in a range lower than 0.01–1.4 (but mostly lower than the detection limit) for popular antibiotics (Boxall et al. 2006a). The highest BCFs (ratio of the analyte concentration detected in the plant tissue to the spiked concentration in the growth medium) have been obtained for carbamazepine (100 L^{-1} kg), fluoxetine and phenytoin (anti-epileptic drugs) in hydroponic experiments, so not for pharmaceuticals used for animal treatment (Wu et al. 2015), but in soils irrigated by treated wastewater. The bioaccumulation factor (BAF) varies between 0 and 4.75 (R = CV/CS; CV, antibiotic residues in vegetables; CV, antibiotic residues in soil) for veterinary antibiotics in vegetables grown in northern China (Hu et al. 2010b).

It is certain that the main mechanism of VPs intake by plants is transpiration-i.e. a passive mechanism (Dodgen et al. 2015; Liu et al. 2013a). The key role in this process is played by the properties of the chemical compounds. In the case of pesticides, the vital role of uptake by plants is fulfilled by hydrophobicity (Inoue et al. 1998; Trapp 2000). The ionic form of compounds also plays a role, but has an opposite effect to the uptake-i.e. ionizable compounds have a limited uptake (Wu et al. 2013). VPs constitute a very wide range of chemical compounds; so, their interaction with plants is assumed to be structure-dependent. Assays of OTC,

CLC and NOR uptake by rice show that maximum concentrations are compound-dependent linear functions of initial soil/water concentrations, and are not related to K_{ow} (Hawker et al. 2013), and this suggests that root intake is connected with the ionization state of these zwitterionic compounds. MQ_S with a large size (>500) have been reported in many studies to not be taken up by plants (Boxall et al. 2006a; Jones-Lepp et al. 2010; Kang et al. 2013; Kumar et al. 2005; Pan et al. 2014). The most probable reason for this is their excessive mass for passive intake and the lack of protein-mediated or energy-dependent activity (Miller et al. 2015). Molecular weight and log K_{ow} have been reported to play a direct role in the uptake of small pharmaceuticals by lettuce (Calderón-Preciado et al. 2012). Given that all VPs are ionized in normal natural conditions, it can be suggested that their uptake will be limited. In addition, ionizable compounds are normally sorbed by soil minerals. The transport of hydrophobic pharmaceuticals is limited and they remain in roots (Dodgen et al. 2015). It has also been reported that basic compounds have higher transport (to leaves) rates using transpiration than acidic or neutral species (Dodgen et al. 2015). Some studies have reported that the intake of small molecules is connected with translocation by water mass flow, and thus pharmaceuticals are concentrated in older leaves (Shenker et al. 2011). However, this is not a rule for all plants. In the case of OTC uptake into alfalfa plants, some energy-dependent processes play a key role (Kong et al. 2007). Moreover, the uptake of this VP is positively related with the pH of the soil solution, where OTC is mostly in a neutral form. All of this information leads to confusion in terms of the prediction of VPs uptake, when the structure of a compound is taken as the predominant factor.

In most cases, hydroponic experiments have been used to estimate the priority of compounds and plant uptake, but the results can be totally different using real soil, because of VPs sorption into organic matter (Wu et al. 2015). The bioconcentrations obtained for pharmaceuticals determined in soil experiments are several times smaller than those for hydroponic experiments, which indisputably confirms that the sorption and transformation of VPs in a real environment significantly decreases their uptake to plants. The high divergence of results from hydroponic and field studies is not a surprise. Natural soil is rich in organic matter, and the correlation between the sorption of pharmaceuticals and the percentage of organic matter is positive (Białk-Bielińska et al. 2012). In the case of carbamazepine, a human pharmaceutical, the plant intake is much lower in the case of organic-rich soil (Shenker et al. 2011). Given that the irrigation of crops with manure increases the organic content in soil, the presumption is that this decreases plant uptake. A study carried out in Canada suggests that the potential for micropollutant uptake into crops under normal farming conditions is low (Sabourin et al. 2012). This has also been proven in a greenhouse study of the uptake of TMP and SMX by cabbage-in a soil rich in organic matter a lack of uptake was observed (Holling et al. 2012). In field studies, real samples presented no (Jones-Lepp et al. 2010) to low uptake values in the order of nanograms per gram in plant tissues (Calderón-Preciado et al. 2011, 2012).

The distribution of antibiotics in a plants taken from China was, in sequence, leaf > stem > root for the edible plants radish, rape, celery, and coriander (Hu et al.

2010b). In contrast, the concentration of SMX and TMP is higher in the root than the leaf (Herklotz et al. 2010). The same has been reported for CIP and NR in tests with barley, carrot, wheat and other forage crop plants (Eggen et al. 2011). The uptake of DC by radish and pakchoi is undetectable, but radish accumulates more SMC and NOR than pakchoi (Wang et al. 2016). OTC, NOR and CTC were found in rice roots in pot tests, but translocation into upper tissues was not observed (Hawker et al. 2013). This agrees with the results of TC_S and NOR tests with rice and soybean plants (Boonsaner and Hawker 2012, 2010). Generally, information for VPs is scarce. More information can be found for human pharmaceuticals (Wu et al. 2013).

5.2 *Metabolism*

The metabolism of small compounds, e.g. pesticides in plants, is generally a multi-step process, including oxidation, reduction, hydrolysis and conjugation (Hoagland et al. 2000). The issue of the metabolism of pharmaceuticals in plants has not been discussed. Individual papers are focused on single compounds, but mostly not veterinary pharmaceuticals. For example, Huber et al. (2012) investigated the metabolism of diclofenac in barley and horse radish, and reported similar activation to that found in mammalian cells in a phase I reaction resulting in the hydroxylated metabolite 4OH-diclofenac, which is conjugated subsequently in phase II to a glucopyranoside, a typical plant-specific metabolite. In the case of triclosan (a bacteriostatic agent used for human purposes), phase I of metabolism was not reported for a carrot, while several conjugates (saccharides, disaccharides, malonic acid, and sulfate) were detected (MacHerius et al. 2012). A study of paracetamol uptake by hydroponically growing Indian mustard (*Brassica juncea* L. Czern.) showed the presence of glutathionyl and a glycoside conjugate, which is similar to the mammalian detoxification pathway (Bartha et al. 2010). Moreover, the uptake of this human pharmaceutical was coupled with an oxidative stress response in the plants. The biotransformation of BNZ_S in reed (*Phragmites australis*) results in the formation of glucosyl-glucosides, acetylglucosides and xylosylglucosides (Podlipná et al. 2013), which are less toxic than native pharmaceuticals, and this can support the use of reed as a tool for phytoremediation.

The accumulation of OTC and NOR antibiotics by soybean results in little effect on the growth rate and maximum levels in plants are observed after 2 days exposure, followed by declining concentrations (Boonsaner and Hawker 2010). This is connected with a high level of degradation for the tested antibiotics in soil (half-life <10 h for both), but it can also be proof that these compounds are degraded in plant cells.

5.3 Toxicity

The toxicity of VPs toward plants is rarely investigated, but these reviews have already been presented, covering both plants' lives in water (e.g. *Lemna minor*) and soil (Bártíková et al. 2016; Carvalho et al. 2014). Most investigations have focused on plant growth and development (Eggen et al. 2011; Furtula et al. 2012; Hoagland 1996; Kołodziejska et al. 2013; Kong et al. 2007; Pomati et al. 2004), chlorophyll content (Opris et al. 2013; Robinson et al. 2005) and seed germination (Liu et al. 2009). Other effects, such as root activity, phosphorus assimilation reduction, root alterations, and reproduction rate changes are also the endpoints of pharmaceutical toxicity toward plants. The toxicity of pharmaceuticals toward crop plants varies significantly between compounds, plant species and test conditions (Carvalho et al. 2014; Chen et al. 2016). The phytotoxicity of VPs has been reported in in vitro assays rather than in soil conditions. Moreover, the concentrations used did not meet the conditions of the environment (Jjemba 2002).

The inhabitation of root elongation (germination) is one of the most important endpoints of the toxicity of veterinary antibiotics (Pan and Chu 2016b). The EC50 of TC, SMT, NOR, ERY and CAP were 14.4, 157, 49.4, 68.8, and 204 mg L^{-1}, respectively. A quantitative structure-activity relationship (QSAR) model showed that in this test kit the hydrophobicity was the most important factor of phytotoxicity. Some studies have reported that the uptake of a human pharmaceutical (carbamazepine) affects nutrient and hormone homeostases in zucchini (*Cucurbita pepo*) plants when a high concentration (up to 4 mg kg^{-1} of soil) is obtained in tests (Carter et al. 2015). Such effects have never been tested for veterinary antibiotics.

Migliore et al. (2003) investigated phytotoxicity and the uptake of ENR in crop plants-*Cucumis sativus, Lactuca sativa, Phaseolus vulgaris* and *Raphanus sativus*. The toxic effect was reported at the high concentration of ENR 5000 μg L^{-1}, while hormesis (increased growth) at the concentrations of 10 and 50 μg L^{-1} of ENR. Interestingly, plants containing ENR at high tissue concentrations can partially metabolize this to CIP, which is normal for animal metabolisms. In an assay of OFN and LIN (together with human pharmaceuticals) uptake and effect on *Eruca sativa L.* and *Zea mays L.* (corn) plants, the use of a low ng L^{-1} of target VPs showed no negative impact on root length or seed germination (Marsoni et al. 2014).

CIP and NR produced negative effects on the growth and development of carrots roots when grown in soil concentrations of 6–10 mg kg^{-1} dry weight (pot tests) (Eggen et al. 2011), which suggests a negative impact on their metabolism in high concentrations of VPs. The same has been stated for SMX, when radish and pakchoi cultivation was irrigated by manure spiked with VPs (Wang et al. 2016).

Seeds seem to be more sensitive to pharmaceuticals than older plants. Values for NOEC (No-Observed Effect Concentration), LOEC (Lowest-Observed Effect Concentration) and EC50 for CTC for cucumber seeds (root length as endpoint) were 0.1, 1 and 48 mg L^{-1} (water used for irrigation), while the same values for a cucumber plant were 100, 300, and >300 mg L^{-1}, respectively (Liu et al. 2009).

Higher toxicity was observed for PEI_S. The EC50 of SAL observed for a *Brassica rapa* was between 1.38 and 3.71 mg kg^{-1} of soil depending on which growth endpoint was tested. The other PEI_S, MON, has already been proven to have herbicidal properties (Hoagland 1996). Two SA_S, SDM and SMT (11.5 mg L^{-1} of nutrient solution), have been proven to affect the root apparatus of barley, while the photosynthetic tissues remained almost unaffected (Michelini et al. 2013). In addition, increased potassium release was noticed, most probably because of an impairment of membrane permeability. In contrast to antibiotics, antiparasitic and anthelmintic pharmaceuticals seem not to be toxic for plants (Kołodziejska et al. 2013; Moore and Kröger 2010; Wagil et al. 2015).

Despite the occurrence of plant uptake of VPs, this uptake usually represents a low percentage of the mass depleted from the systems. Phytoremediation also depends on the rhizospheric microbial communities, which in natural conditions are rich and have a meaningful role in the elimination of pharmaceuticals. Considering the concentration levels that pharmaceuticals normally present in environmental matrices, either in wastewaters or reclaimed water, it is not expected that phytotoxic effects will occur in phytoremediation designed systems (Carvalho et al. 2014). In the case of manure, where the VPs concentration can be higher, an effect lethal to plants is also unlikely. It should be added that pharmaceuticals in soils can affect not only plants and bacteria, but also earthworms (Pino et al. 2015). The effect of antibiotics on a rhizosphere (microorganism community near the roots) is under investigation and this effect seems to be unlike that in the bulk of soil (Jechalke et al. 2014).

The results of studies focused on the prediction models of pharmaceutical concentrations in the environment indicate that the contamination of TC_S in top soil may represent a major issue both for compliance with maximum residue levels in food (100–300 ng g^{-1}) and for the worst-case scenario claims made for organic products (Brambilla and Testa 2014). The most critical compounds are five antibiotics and two analgesic anti-inflammatories (Verlicchi and Zambello 2014). Considering the environmental levels (much lower than concentrations in tests) and the fate of antibiotics in soil (sorption, degradation and chelating with metals), a low level of toxicity from VPs toward plants on real fields is expected (Liu et al. 2009). Given the acceptable daily intake for popular antibiotics and the concentration found in edible plants, there is little evidence of an appreciable risk for human (Aryal and Reinhold 2011; Boxall et al. 2006a; Kang et al. 2013). This was also stated by Prosser and Sibley, who combined information about the lowest therapeutic dose (mg day^{-1}), and the no observable adverse effect level (mg^{-1} kg^{-1} day^{-1}), and calculated the safety factor and the acceptable daily intake of pharmaceuticals (Prosser and Sibley 2015).

6 Current Legislation and Regulation

The public health hazards related to the use of VPs in agriculture and aquaculture involve several problems such as the increased risk of developing allergies in individuals with hyper-sensitivity or the development of antibiotic-resistant bacteria, because of the transmission of antibiotic-resistant pathogenic bacteria through the food chain (Botsoglou and Fletouirs 2001; Dibner and Richards 2005; Institute of Medicine 1989; Motarjemi et al. 2014; Pruneda 1950; WHO 2000). Moreover, the globalization of the food supply requires the harmonization of polices and standards based on a common understanding of food safety among authorities in countries around the world. Given the reported use in the EU in 2004 and animal production data, the average VP_S consumption factor (VPCFa) for antibiotics was 141 mg kg^{-1} meat, and for BNZ_S 6 mg kg^{-1} meat (Kools et al. 2008). To protect consumers from adverse health effects caused by food-borne residues of VPs, many countries maintain strict controls on the authorization, labeling, and use of VPs in food-producing animals. The European Union (EU), Food and Agriculture Organization (FAO) and Food and Drug Administration (FDA) have established certain regulations to monitor the pharmaceutical residues in foods of animal origin (Botsoglou and Fletouirs 2001). Some compounds have been banned for food safety reasons; for other agents the FDA is implementing a voluntary plan with industry to phase out or discourage their use because antimicrobial resistance may not be completely preventable (US Food and Drug Administration 2013). This approach can also be observed in EU countries. In the Euro Zone, in order to reduce human exposure to antibiotic residues, the consumption of any antibiotic is prohibited for growth promotion purposes (The Commision of the European Communities 2005). However, the evaluation of the exposure to pharmaceutical residues has always been controversial, particularly in those geographical regions which do not follow restrictive regulations on the use of veterinary medicines.

As mentioned, VPs may be excreted directly to soils by pasture animals or they may be released into the soil environment when animal manure is applied to land. Some of these are used in fish farms and are introduced directly into the aquatic system (e.g. Beausse 2004; Carbonell et al. 2009; Kuster et al. 2004; Zhao et al. 2010). In some cases, VP_S products are administrated in such a way that user exposure is more likely, or more extensive than with human counterparts. Among examples of this are the vaccination of poultry and dipping of sheep for ectoparasitic conditions (Woodward 2008). Moreover, many human medicinal products contain the same active ingredients as their veterinary counterparts. These pharmaceuticals are released into municipal sewage systems, and it is well-known that, depending on their chemical structure, they can survive passage through sewage treatment plants. Many of them may also show a strong tendency to sorption in sewage sludge, which is used to some extent for agricultural purposes, so this is another way of their introduction into soil (Buchberger 2011). It has been established that approximately ten million dry tons of digested municipal sludge (known as biosolids) is produced each year in the EU (about 40% of it is used in agriculture), and seven million dry tons in the U.S. (about 50% for agricultural purposes) (Dodgen et al. 2014; Environmental Protection Agency (2010); European Commission 2002; Roccaro and Vagliasindi 2014). European legislation (European Commission 2004) permits its use when concentrations of metals in soil do not exceed the maximum permissible limits; however, the concentrations of VPs in the soil (sludge) are not regulated (Carbonell et al. 2009; Clarke and Cummins 2015; European Commission 2004; Roccaro and Vagliasindi 2014). Moreover, according to the Urban Wastewater Treatment Directive (UWWTD), the 28 European Union (EU-28) members are required to collect and treat their urban wastewater. The reuse of the sludge is also encouraged and its final disposal to surface waters has been banned (Kelessidis and Stasinakis 2012). Treated wastewater water reuse is also growing in the U.S. (by 15% a year) (Wade Miller 2006); in many cases it is used for agricultural and landscape irrigation. It is clear that the use of treated wastewater is more regulated than the use of manure, while both can be contaminated by pharmaceuticals.

Taking into account the presented information, the presence of many VP_S at concentrations of up to the low mg kg^{-1} level in agricultural soils is well-confirmed and obvious and might cause adverse effects on the ecosystem and human health (Białk-Bielińska et al. 2016; García-Santiago et al. 2016; Rodriguez-Ruiz et al. 2015; Tarazona et al. 2013; Tuhkanen et al. 2007). However, as mentioned, the concentration limits of VPs in the environment (also in the soil compartments) are still not regulated, even though growing concern in the U.S. and Europe has resulted in the prescription of environmental risk assessments of VP_S (European Medicines Agency 2004, 2008, 2009; The European Agency for the Evaluation of Medical Products 2000; World Health Organization 2001). For example, Montforts (1999) has provided detailed emission and distribution models as well as environmental risk assessment for VPs; and regulatory demands on data quality for the environmental risk assessment of pharmaceuticals have been presented by Küster et al. (2009), whereas relevant EU legislation and risk assessment strategies for the control of emerging contaminants (e.g. pharmaceuticals) resulting from the application of biosolids to agricultural lands have been elaborated by Clarke and Cummins (2015). A risk assessment of persistent pharmaceuticals in biosolids taking into account uncertainty, has been presented by García-Santiago et al. (2016). There is no EU legislation defining the permissible levels of antibiotic concentrations in soils. European Agency for the Evaluation of Medicinal Products sets a threshold value of 0.1 mg kg^{-1} for residues of VPs in soils (EMEA 1996); however, this threshold only applies to approving new substances (Martínez-Carballo et al. 2007). In addition, there is no regulation of VPs content in manure to regulate its use as a fertilizer of crop fields.

7 Conclusion and Summary

It is evident that VPs are observed in the agricultural environment because of the intensive use of veterinary medicines for animal production and the subsequent manure application onto crop fields. The concentrations found in manure (Table 2) are as high as mg kg^{-1}, while they are lower in soil (mostly μg kg^{-1} or ng kg^{-1}). What is a noticeable is that VPs occur in a thin layer situated on the surface (up to ~40 cm depth), which is rich in organic matter with sorption properties. The sorption process can be reversible, and provide long-term evidence of VPs in soil media. However, the sorption properties decrease the negative effect of VPs. This was confirmed in the research cited in this review.

Firstly, sorption can effectively prevent leaching to groundwater, which has been suggested to pose a risk to human health. A strong interaction with soil particles has been determined for TC_S and FQ_S, whereas relatively weak interactions are exhibited by e.g. SA_S and IMD_S (Białk-Bielińska et al. 2012; Song and Guo 2014). The leaching of antibiotics and antiparasitic pharmaceuticals into groundwater is observed in laboratory tests, while they are not determined in field screening. The most probable reason for this is that laboratory tests give overestimated data, because concentration levels used in sorption studies (mg kg^{-1}) are often significantly higher than those that would normally be present in the environment (ng kg^{-1}). While sorption is a concentration-dependent process, new tools for leaching need to be investigated [e.g. the concentration-dependent effective distribution coefficient (K_d^{eff}) (Srinivasan et al. 2014)]. Moreover, leaching models used seem to underestimate the leaching behavior of VPs, probably due to a lack of consideration of factors affecting leaching in the field [e.g. degradation process, temperature changes, bioavailability (Blackwell et al. 2009)].

Secondly, a number of studies have shown that sorption may lead to the reduction of antibiotic effects on soil microbes (Hund-Rinke et al. 2004; Kotzerke et al. 2008; Thiele-Bruhn 2005). Numerous studies have documented changes in soil microbial community structure as a result of exposure to antibiotics, while the results reported on wetlands disagree (Berglund et al. 2014). Experiments have proved that the ratios of bacteria/fungi and Gram positive/Gram negative bacteria can be changed when mg/kg of VPs in soil are applied. Long-term studies frequently detect a recovery of the microbe community biomass and growth that was initially inhibited (Demoling et al. 2009; Kotzerke et al. 2008). Land application of manure from antibiotic-treated animals is reported to frequently increase the abundance of resistant bacteria and antibiotic resistance genes in soils (Ghosh and LaPara 2007; Martínez 2008). Currently, the development of resistance is not believed to be promoted by continuing exposure to sublethal concentrations of antibiotics, but it appears to be more important that resistant microorganisms are introduced together with manure. The resistance of bacteria obtained in soil may also be due to naturally occurring antibiotics and resistance genes.

Thirdly, the sorption of VPs plays an important role in plant uptake. Information about the intake of VPs by plants is mostly limited to hydroponic conditions, while the bioconcentration obtained for pharmaceuticals determined in soil experiments is several times smaller. The analysis of real samples taken from farmland show the absence of VPs in plant tissues (Wu et al. 2015). Given that the irrigation of crops with manure increases the organic content in soil, the presumption is that this decreases plant uptake. Several studies have suggested that the potential for VPs uptake into crops under normal farming conditions is low (Calderón-Preciado et al. 2011, 2012; Holling et al. 2012; Jones-Lepp et al. 2010; Sabourin et al. 2012). MQ_S with a large size have been reported in many studies to not be taken up by crops (Boxall et al. 2006a; Jones-Lepp et al. 2010; Kang et al. 2013; Kumar et al. 2005; Pan et al. 2014). In the case of smaller VPs, the ionization state seems to limit the root uptake and further translocation in plant tissue. Moreover, it seems that plants have their own defenses against pharmaceuticals and their metabolism tends in the direction of conjugate production. The phytotoxicity of VPs has been reported in in vitro assays rather than in soil conditions. Plants living in constructed wetlands can exist for years without any visible changes caused by pharmaceuticals (Verlicchi and Zambello 2014).

To sum up, the application of manure on crop fields poses risks, but this is substantially limited by the sorption of VPs into soil particles and the natural occurring processes of biotic and abiotic degradation. The composting of manure before use in aerobic conditions significantly reduces the bulk of VPs, and currently this is the only way eliminate antibiotics in manure before use as fertilizer, as VPs cannot be eliminated in animal production. The application of manure is not a continuous process, but occurs periodically before crop seeding. The microorganisms living in soil are generally not impacted upon by VPs in the concentrations observed in soils. If a negative impact occurs, the bacterial community has the ability to resist antibiotics and reconstruct the population. The plant uptake of VPs is negligible and limited to the roots. In the summer period, the degradation of VPs is accelerated, and the half-life of some antibiotics in manure and soil is in a matter of days. The higher half-life of some VPs is connected with the fact that sorbed fractions of compounds are much less susceptible to both biotic and abiotic degradation. The process of fertilization via the application of manure is of agricultural interest due mainly to its organic matter concentrations and nutrient input, and is much safer than the irrigation of fields with treated wastewater or wastewater sludge. Wastewater products are rich not only in antibiotics, but also human pharmaceuticals, metals, alkyl-phenols, phthalates and other micropollutants. Moreover, the continuous process of irrigation and the consequential accumulation of the above compounds can pose risks for not only organisms living there, but also human health.

The current state of the art in VPs in manure and their agricultural use is much broader than in 2003, when Boxall and coworkers investigated the risk caused by veterinary medicines in the environment (Boxall et al. 2003). Up-to-date information concerning the presence of antibiotics and the related presence of veterinary compounds, degradation, uptake by plants, toxicity to microorganisms, leaching and connected issues is presented in this review. Tentatively, it can be stated that VPs in manure do not pose a risk for the agricultural environment. However, we have to bear in mind that VPs are a diverse group of substances, and each of them should be approached separately. For example, the SA_S show a potential for leaching and a longer life time in soil. The one concern is that most current information comes from experiments with short durations, high concentrations of target substances and conditions divergent to the environment (e.g. hydroponic experiments into uptake). Moreover, only rare reports have focused on degradation/ transformation products, which are generally more labile and can affect the environment in a different manner to the native VPs. There is also a lack of experiments on mixtures of VPs and other pollutants.

Acknowledgements Financial support was provided by the National Science Center (Poland) under decision number UMO-2014/13/N/ST4/04127, by the National Center for Research and Development (NCBR) (Poland) under grant TANGO1/268806/NCBR/2015 and by the Polish Ministry of Research and Higher Education under grant DS 530–8616-D593–16 and DS 530–8615-D592–15.

References

Allen HK, Donato J, Wang HH, Cloud-Hansen KA, Davies J, Handelsman J (2010) Call of the wild: antibiotic resistance genes in natural environments. Nat Rev Microbiol Rev 8:251–259. doi:10.1038/nrmicro2312

Allison SD, Martiny JBH (2008) Resistance, resilience, and redundancy in microbial communities. Proc Natl Acad Sci U S A 105:11512–11519. doi:10.1073/pnas.0801925105

Anderson CR, Rupp HS, Wu WH (2005) Complexities in tetracycline analysis—chemistry, matrix extraction, cleanup, and liquid chromatography. J Chromatogr A 1075:23–32. doi:10.1016/j.chroma.2005.04.013

Andreu V, Vazquez-Roig P, Blasco C, Picó Y (2009) Determination of tetracycline residues in soil by pressurized liquid extraction and liquid chromatography tandem mass spectrometry. Anal Bioanal Chem 394:1329–1339. doi:10.1007/s00216-009-2635-x

Angenent LT, Mau M, George U, Zahn JA, Raskin L (2008) Effect of the presence of the antimicrobial tylosin in swine waste on anaerobic treatment. Water Res 42:2377–2384. doi:10.1016/j.watres.2008.01.005

Arikan OA, Sikora LJ, Mulbry W, Khan SU, Rice C, Foster GD (2006) The fate and effect of oxytetracycline during the anaerobic digestion of manure from therapeutically treated calves. Process Biochem 41:1637–1643. doi:10.1016/j.procbio.2006.03.010

Arikan OA, Sikora LJ, Mulbry W, Khan SU, Foster GD (2007) Composting rapidly reduces levels of extractable oxytetracycline in manure from therapeutically treated beef calves. Bioresour Technol 98:169–176. doi:10.1016/j.biortech.2005.10.041

Aryal N, Reinhold DM (2011) Phytoaccumulation of antimicrobials from biosolids: impacts on environmental fate and relevance to human exposure. Water Res 45:5545–5552. doi:10.1016/j.watres.2011.08.027

Aust MOO, Godlinski F, Travis GR, Hao X, McAllister TA, Leinweber P, Thiele-Bruhn S (2008) Distribution of sulfamethazine, chlortetracycline and tylosin in manure and soil of Canadian feedlots after subtherapeutic use in cattle. Environ Pollut 156:1243–1251. doi:10.1016/j.envpol.2008.03.011

Babić S, Periša M, Škorić I (2013) Photolytic degradation of norfloxacin, enrofloxacin and ciprofloxacin in various aqueous media. Chemosphere 91:1635–1642. doi:10.1016/j.chemosphere.2012.12.072

Baran W, Sochacka J, Wardas W (2006) Toxicity and biodegradability of sulfonamides and products of their photocatalytic degradation in aqueous solutions. Chemosphere 65:1295–1299. doi:10.1016/j.chemosphere.2006.04.040

Barnes KK, Kolpin DW, Furlong ET, Zaugg SD, Meyer MT, Barber LB (2008) A national reconnaissance of pharmaceuticals and other organic wastewater contaminants in the United States—(I) groundwater. Sci Total Environ 402:192–200. doi:10.1016/j.scitotenv.2008.04.028

Barra Caracciolo A, Topp E, Grenni P (2015) Pharmaceuticals in the environment: biodegradation and effects on natural microbial communities. A review. J Pharm Biomed Anal 106:25–36. doi:10.1016/j.jpba.2014.11.040

Bartelt-Hunt S, Snow DD, Damon-Powell T, Miesbach D (2011) Occurrence of steroid hormones and antibiotics in shallow groundwater impacted by livestock waste control facilities. J Contam Hydrol 123:94–103. doi:10.1016/j.jconhyd.2010.12.010

Bartha B, Huber C, Harpaintner R, Schröder P (2010) Effects of acetaminophen in Brassica juncea L. Czern.: investigation of uptake, translocation, detoxification, and the induced defense pathways. Environ Sci Pollut Res 17:1553–1562. doi:10.1007/s11356-010-0342-y

Bártíková H, Podlipná R, Skálová L (2016) Veterinary drugs in the environment and their toxicity to plants. Chemosphere 144:2290–2301. doi:10.1016/j.chemosphere.2015.10.137

Batt AL, Snow DD, Aga DS (2006) Occurrence of sulfonamide antimicrobials in private water wells in Washington County, Idaho, USA. Chemosphere 64:1963–1971. doi:10.1016/j.chemosphere.2006.01.029

Baynes RE, Dedonder K, Kissell L, Mzyk D, Marmulak T, Smith G, Tell L, Gehring R, Davis J, Riviere JE (2016) Health concerns and management of select veterinary drug residues. Food Chem Toxicol 88:112–122. doi:10.1016/j.fct.2015.12.020

Beausse J (2004) Selected drugs in solid matrices: a review of environmental determination, occurrence and properties of principal substances. Trends Anal Chem 23:753–761. doi:10.1016/j.trac.2004.08.005

Berg J, Thorsen MK, Holm PE, Jensen J, Nybroe O, Brandt KK (2010) Cu exposure under field conditions coselects for antibiotic resistance as determined by a novel cultivation-independent bacterial community tolerance assay. Environ Sci Technol 44:8724–8728. doi:10.1021/es101798r

Berglund B, Khan GA, Weisner SEB, Ehde PM, Fick J, Lindgren P-E (2014) Efficient removal of antibiotics in surface-flow constructed wetlands, with no observed impact on antibiotic resistance genes. Sci Total Environ 476–477:29–37. doi:10.1016/j.scitotenv.2013.12.128

Białk-Bielińska A, Kumirska J, Palavinskas R, Stepnowski P (2009) Optimization of multiple reaction monitoring mode for the trace analysis of veterinary sulfonamides by LC-MS/MS. Talanta 80:947–953. doi:10.1016/j.talanta.2009.08.023

Białk-Bielińska A, Stolte S, Arning J, Uebers U, Böschen A, Stepnowski P, Matzke M (2011) Ecotoxicity evaluation of selected sulfonamides. Chemosphere 85:928–933. doi:10.1016/j.chemosphere.2011.06.058

Białk-Bielińska A, Maszkowska J, Mrozik W, Bielawska A, Kołodziejska M, Palavinskas R, Stepnowski P, Kumirska J (2012) Sulfadimethoxine and sulfaguanidine: their sorption potential on natural soils. Chemosphere 86:1059–1065. doi:10.1016/j.chemosphere.2011.11.058

Białk-Bielińska A, Maszkowska J, Puckowski A, Stepnowski P (2014) Exposure and hazard identification of sulphonamides in the terrestrial environment. In: Environmental risk assessment in soil contamination ISBN 978-953-51-1235-8 InTech. Rijeka Croatia

Białk-Bielińska A, Kumirska J, Borecka M, Caban M, Paszkiewicz M, Pazdro K, Stepnowski P (2016) Selected analytical challenges in the determination of pharmaceuticals in drinking/marine waters and soil/sediment samples. J Pharm Biomed Anal. doi:10.1016/j.jpba.2016.01.016

Binh CTT, Heuer H, Kaupenjohann M, Smalla K (2008) Piggery manure used for soil fertilization is a reservoir for transferable antibiotic resistance plasmids. FEMS Microbiol Ecol 66:25–37. doi:10.1111/j.1574-6941.2008.00526.x

Bissett A, Burke C, Cook PLM, Bowman JP (2007) Bacterial community shifts in organically perturbed sediments. Environ Microbiol 9:46–60. doi:10.1111/j.1462-2920.2006.01110.x
Blackwell PA, Kay P, Boxall ABA (2007) The dissipation and transport of veterinary antibiotics in a sandy loam soil. Chemosphere 67:292–299. doi:10.1016/j.chemosphere.2006.09.095
Blackwell PA, Kay P, Ashauer R, Boxall ABA (2009) Effects of agricultural conditions on the leaching behaviour of veterinary antibiotics in soils. Chemosphere 75:13–19. doi:10.1016/j.chemosphere.2008.11.070
Boleas S, Alonso C, Pro J, Fernández C, Carbonell G, Tarazona JV (2005) Toxicity of the antimicrobial oxytetracycline to soil organisms in a multi-species-soil system (MS·3) and influence of manure co-addition. J Hazard Mater 122:233–241. doi:10.1016/j.jhazmat.2005.03.003
Boonsaner M, Hawker DW (2010) Accumulation of oxytetracycline and norfloxacin from saline soil by soybeans. Sci Total Environ 408:1731–1737. doi:10.1016/j.scitotenv.2009.12.032
Boonsaner M, Hawker DW (2012) Investigation of the mechanism of uptake and accumulation of zwitterionic tetracyclines by rice (Oryza sativa L.). Ecotoxicol Environ Saf 78:142–147. doi:10.1016/j.ecoenv.2011.11.023
Borgman O, Chefetz B (2013) Combined effects of biosolids application and irrigation with reclaimed wastewater on transport of pharmaceutical compounds in arable soils. Water Res 47:3431–3443. doi:10.1016/j.watres.2013.03.045
Botsoglou NA, Fletouirs DJ (2001) Drug residues in food. Pharmacology, food safety and analysis. Marcel Dekker, Inc., New York
Boxall ABA, Kolpin DW, Halling-Sorensen B, Tolls J (2003) Are veterinary medicines causing environmental risk? Environ Sci Technol 37:286A–294A. doi:10.1021/es032519b
Boxall ABA, Fogg LA, Blackwell PA, Kay P, Pemberton EJ, Croxford A (2004) Veterinary medicines in the environment. Rev Environ Contam Toxicol 180:1–91. doi:10.1007/0-387-21729-0_1
Boxall ABA, Fogg L, Baird D, Telfer T, Lewis C, Gravell A, Boucard T (2006a) Targeted monitoring study for veterinary medicines in the environment. Environment Agency, Bristol
Boxall ABA, Johnson P, Smith EJ, Sinclair CJ, Stutt E, Levy LS (2006b) Uptake of veterinary medicines from soils into plants. J Agric Food Chem 54:2288–2297. doi:10.1021/jf053041t
Boxall, ABA, Tiede, K, Bevan, R, Tam, C, Levy, S, (2006) Desk-based study of current knowledge on veterinary medicines in drinking water and estimation of potential levels. The Food and Environment Research Agency, Sand Hutton
Brambilla G, Testa C (2014) Food safety/food security aspects related to the environmental release of pharmaceuticals. Chemosphere 115:81–87. doi:10.1016/j.chemosphere.2014.01.024
Brandt KK, Sjøholm OR, Krogh KA, Halling-Sørensen B, Nybroe O (2009) Increased pollution-induced bacterial community tolerance to sulfadiazine in soil hotspots amended with artificial root exudates. Environ Sci Technol 43:2963–2968. doi:10.1021/es803546y
Buchberger W (2011) Current approaches to trace analysis of pharmaceuticals and personal care products in the environment. J Chromatogr A 1218:603–618. doi:10.1016/j.chroma.2010.10.081
Bundt M, Widmer F, Pesaro M, Zeyer J, Blaser P (2001) Preferential flow paths: biological "hot spots" in soils. Soil Biol Biochem 33:729–738. doi:10.1016/S0038-0717(00)00218-2
Byrne-Bailey KG, Gaze WH, Kay P, Boxall ABA, Hawkey PM, Wellington EMH (2009) Prevalence of sulfonamide resistance genes in bacterial isolates from manured agricultural soils and pig slurry in the United Kingdom. Antimicrob Agents Chemother 53:696–702. doi:10.1128/AAC.00652-07
Calderón-Preciado D, Jimenez-Cartagena C, Matamoros V, Bayona JM (2011) Screening of 47 organic microcontaminants in agricultural irrigation waters and their soil loading. Water Res 45:221–231. doi:10.1016/j.watres.2010.07.050
Calderón-Preciado D, Renault Q, Matamoros V, Cañameras N, Bayona JM (2012) Uptake of organic emergent contaminants in spath and lettuce: an in vitro experiment. J Agric Food Chem 60:2000–2007. doi:10.1021/jf2046224

Campagnolo ER, Johnson KR, Karpati A, Rubin CS, Kolpin DW, Meyer MT, Esteban JE, Currier RW, Smith K, Thu KM, McGeehin M (2002) Antimicrobial residues in animal waste and water resources proximal to large-scale swine and poultry feeding operations. Sci Total Environ 299:89–95. doi:10.1016/S0048-9697(02)00233-4

Campo J (2015) Advanced mass spectrometry for food safety and quality—emerging contaminants, comprehensive analytical chemistry. Elsevier, Amsterdam. doi:10.1016/B978-0-444-63340-8.00010-8

Carbonell G, Pro J, Gomez N, Babin MM, Fernandez C, Alonso E, Tarazona JV (2009) Sewage sludge applied to agricultural soil: ecotoxicological effects on representative soil organisms. Ecotoxicol Environ Saf 72:1309–1319. doi:10.1016/j.ecoenv.2009.01.007

Carlson JC, Mabury SA (2006) Dissipation kinetics and mobility of chlortetracycline, tylosin and monensin in an agricultural soil in Northumberland County, Ontario, Canada. Environ Toxicol Chem 25:1–10. doi:10.1897/04-657R.1

Carter LJ, Harris E, Williams M, Ryan JJ, Kookana RS, Boxall ABA (2014) Fate and uptake of pharmaceuticals in soil—plant systems. J Agric Food Chem 62:816–825. doi:10.1021/jf404282y

Carter LJ, Williams M, Böttcher C, Kookana RS (2015) Uptake of pharmaceuticals influences plant development and affects nutrient and hormone homeostases. Environ Sci Technol 49:12509–12518. doi:10.1021/acs.est.5b03468

Carvalho PN, Basto MCP, Almeida CMR, Brix H (2014) A review of plant-pharmaceutical interactions: from uptake and effects in crop plants to phytoremediation in constructed wetlands. Environ Sci Pollut Res 21(20):11729–11763. doi:10.1007/s11356-014-2550-3

Chee-Sanford JC, Mackie RI, Koike S, Krapac IG, Lin YF, Yannarell AC, Maxwell S, Aminov RI (2009) Fate and transport of antibiotic residues and antibiotic resistance genes following land application of manure waste. J Environ Qual 38:1086–1108. doi:10.2134/jeq2008.0128

Chen Y, Hong L, Zongping W, Huijie L, Tao T, Zuo Y (2012) Photodegradation of selected β-blockers in aqueous fulvic acid solutions: kinetics, mechanism and product analysis. Water Res 46:2965–2872. doi:10.1016/j.watres.2012.03.025

Chen Y, Liang Q, Zhou D, Wang Z, Tao T, Zuo Y (2013) Photodegradation kinetics, products and mechanism of timolol under simulated sunlight. J Hazardous Mater 252–253:220–226. doi:10.1016/j.jhazmat.2013.02.035

Chen C, Li J, Chen P, Ding R, Zhang P, Li X (2014) Occurrence of antibiotics and antibiotic resistances in soils from wastewater irrigation areas in Beijing and Tianjin, China. Environ Pollut 193:94–101. doi:10.1016/j.envpol.2014.06.005

Chen J, Xu H, Sun Y, Huang L, Zhang P, Zou C, Yu B, Zhu G, Zhao C (2016) Interspecific differences in growth response and tolerance to the antibiotic sulfadiazine in ten clonal wetland plants in South China. Sci Total Environ 543:197–205. doi:10.1016/j.scitotenv.2015.11.015

Chenxi W, Spongberg AL, Witter JD (2008) Determination of the persistence of pharmaceuticals in biosolids using liquid-chromatography tandem mass spectrometry. Chemosphere 73:511–518. doi:10.1016/j.chemosphere.2008.06.026

Christian T, Schneider RJ, Färber HA, Skutlarek D, Meyer MT, Goldbach HE (2003) Determination of antibiotic residues in manure, soil, and surface waters. Acta Hydrochim Hydrobiol 31:36–44. doi:10.1002/aheh.200390014

Chuang YH, Zhang Y, Zhang W, Boyd SA, Li H (2015) Comparison of accelerated solvent extraction and quick, easy, cheap, effective, rugged and safe method for extraction and determination of pharmaceuticals in vegetables. J Chromatogr A 1404:1–9. doi:10.1016/j.chroma.2015.05.022

Clarke RM, Cummins E (2015) Evaluation of "classic" and emerging contaminants resulting from the application of biosolids to agricultural lands: a review. Hum Ecol Risk Assess 21:492–513. doi:10.1080/10807039.2014.930295

Clervil E, Usman M, Emmanuel E, Chatain V, Hanna K (2013) Sorption of nalidixic acid onto sediments under batch and dynamic flow conditions. Chem Geol 335:63–74. doi:10.1016/j.chemgeo.2012.10.041

Colinas C, Ingham E, Molina R (1994) Population responses of target and non-target forest soil organisms to selected biocides. Soil Biol Biochem 26:41–47. doi:10.1016/0038-0717(94)90193-7

Conkle JL, Lattao C, White JR, Cook RL (2010) Competitive sorption and desorption behavior for three fluoroquinolone antibiotics in a wastewater treatment wetland soil. Chemosphere 80:1353–1359. doi:10.1016/j.chemosphere.2010.06.012

Cortés JM, Larsson E, Jönsson JÅ (2013) Study of the uptake of non-steroid anti-inflammatory drugs in wheat and soybean after application of sewage sludge as a fertilizer. Sci Total Environ 449:385–389. doi:10.1016/j.scitotenv.2013.01.061

Cundliffe E. (1989) How antibiotic-producing organisms avoid suicide. Annual Review of Microbiology 43: 207-233. doi: 10.1146/annurev.mi.43.100189.001231

Daghrir R, Drogui P (2013) Tetracycline antibiotics in the environment: a review. Environ Chem Lett 11:209–227. doi:10.1007/s10311-013-0404-8

Davis JG, Truman CC, Kim SC, Ascough JC, Carlson K (2006) Antibiotic transport via runoff and soil loss. J Environ Qual 35:2250–2260. doi:10.2134/jeq2005.0348

Delay M, Lager T, Schulz HD, Frimmel FH (2007) Comparison of leaching tests to determine and quantify the release of inorganic contaminants in demolition waste. Waste Manag 27:248–255. doi:10.1016/j.wasman.2006.01.013

Demoling LA, Bååth E (2008) No long-term persistence of bacterial pollution-induced community tolerance in tylosin-polluted soil. Environ Sci Technol 42:6917–6921. doi:10.1021/es8004706

Demoling LA, Bååth E, Greve G, Wouterse M, Schmitt H (2009) Effects of sulfamethoxazole on soil microbial communities after adding substrate. Soil Biol Biochem 41:840–848. doi:10.1016/j.soilbio.2009.02.001

Devillers J (2009) Ecotoxicology modeling. Springer, Berlin. doi:10.1007/978-1-4419-0197-2

Dghrir A, Drogui P (2013) Tetracycline antibiotics in the environment: a review. Environ Chem Lett 11:209–227. doi:10.1007/s10311-013-0404-8

Di Guardo A, Finizio A (2016) A moni-modelling approach to manage groundwater risk to pesticide leaching at regional scale. Sci Total Environ 545–546:200–209. doi:10.1016/j.scitotenv.2015.12.056

Díaz-Cruz M (2007) Recent advances in LC-MS residue analysis of veterinary medicines in the terrestrial environment. Trends Anal Chem 26:637–646. doi:10.1016/j.trac.2007.04.004

Diaz-Cruz MS, Garca-Galan MJ, Barceló D (2008) Highly sensitive simultaneous determination of sulfonamide antibiotics and one metabolite in environmental waters by liquid chromatography-quadrupole linear ion trap-mass spectrometry. J Chromatogr A 1193:50–59. doi:10.1016/j.chroma.2008.03.029

Díaz-Cruz MS, López de Alda MJ, Barceló D (2003) Environmental behavior and analysis of veterinary and human drugs in soils, sediments and sludge. Trends Anal Chem 22:340–351. doi:10.1016/S0165-9936(03)00603-4

Dibner JJ, Richards JD (2005) Antibiotic growth promoters in agriculture: history and mode of action. Poult Sci 84:634–643. doi:10.1093/ps/84.4.634

Ding C, He J (2010) Effect of antibiotics in the environment on microbial populations. Appl Microbiol Biotechnol 87:925–941. doi:10.1007/s00253-010-2649-5

Ding Y, Zhang W, Gu C, Xagoraraki I, Li H (2011) Determination of pharmaceuticals in biosolids using accelerated solvent extraction and liquid chromatography/tandem mass spectrometry. J Chromatogr A 1218:10–16. doi:10.1016/j.chroma.2010.10.112

Dodgen LK, Li J, Wu X, Lu Z, Gan JJ (2014) Transformation and removal pathways of four common PPCP/EDCs in soil. Environ Pollut 193:29–36. doi:10.1016/j.envpol.2014.06.002

Dodgen LK, Ueda A, Wu X, Parker DR, Gan J (2015) Effect of transpiration on plant accumulation and translocation of PPCP/EDCs. Environ Pollut 198:144–153. doi:10.1016/j.envpol.2015.01.002

Dolliver H, Gupta S, Noll S (2008) Antibiotic degradation during manure composting. J Environ Qual 37:1245–1253. doi:10.2134/jeq2007.0399

Dorival-García N, Labajo-Recio C, Zafra-Gómez A, Juárez-Jiménez B, Vílchez JL (2015) Improved sample treatment for the determination of 17 strong sorbed quinolone antibiotics from compost by ultra high performance liquid chromatography tandem mass spectrometry. Talanta 138:247–257. doi:10.1016/j.talanta.2015.03.011

Dzierżawski A (2012) Potrzeba racjonalnego stosowania antybiotyków w praktyce weterynaryjnej. Życie Weter 87:316–321

Eggen T, Asp TN, Grave K, Hormazabal V (2011) Uptake and translocation of metformin, ciprofloxacin and narasin in forage- and crop plants. Chemosphere 85:26–33. doi:10.1016/j.chemosphere.2011.06.041

Environmental Protection Agency (2010) A guide to understanding biosolids issues. Santa Rosa, CA

Environmental Risk Assessment for Veterinary Medicinal Products Other than GMO-containing and Immunological Products. EMEA/CVMP/055/96-final, London

Estévez E, Cabrera MC, Molina-Díaz A, Robles-Molina J, Palacios-Díaz Mdel P (2012) Screening of emerging contaminants and priority substances (2008/105/EC) in reclaimed water for irrigation and groundwater in a volcanic aquifer (Gran Canaria, Canary Islands, Spain). Sci Total Environ 433:538–546. doi:10.1016/j.scitotenv.2012.06.031

European Commission (2002) Disposal and recycling routes for sewage sludge, economic analysis. Luxembourg

European Commission (2004) Draft discussion document for the ad hoc meeting on biowaste and sludges. Geneva

European Medicines Agency (2004) Environmental impact assessment for veterinary medicinal products. Phase II guidance. London

European Medicines Agency (2008) Reflection paper on the implementation of directive 2001/82/EC, as amended, in respect to the assessment of environmental risks of veterinary medicinal products 1–15. London

European Medicines Agency (2009) Revised guidiline on environmental impact assessment for veterinary medicinal products, report EMEA/CVMP/ERA/418282/2005 Rev. 1 Corr. London

Fahrenfeld N, Knowlton K, Krometis LA, Hession WC, Xia K, Lipscomb E, Libuit K, Green BL, Pruden A (2014) Effect of manure application on abundance of antibiotic resistance genes and their attenuation rates in soil: Field-scale mass balance approach. Environ Sci Technol 48:2643–2650. doi:10.1021/es404988k

Fang H, Han Y, Yin Y, Pan X, Yu Y (2014) Variations in dissipation rate, microbial function and antibiotic resistance due to repeated introductions of manure containing sulfadiazine and chlortetracycline to soil. Chemosphere 96:51–56. doi:10.1016/j.chemosphere.2013.07.016

Fatta-Kassinos D, Kalavrouziotis IK, Koukoulakis PH, Vasquez MI (2011) The risks associated with wastewater reuse and xenobiotics in the agroecological environment. Sci Total Environ 409:3555–3563. doi:10.1016/j.scitotenv.2010.03.036

Fernandes JP, Almeida CMR, Pereira AC, Ribeiro IL, Reis I, Carvalho P, Basto MCP, Mucha AP (2015) Bioresource technology microbial community dynamics associated with veterinary antibiotics removal in constructed wetlands microcosms. Bioresour Technol 182:26–33. doi:10.1016/j.biortech.2015.01.096

Forsberg KJ, Reyes A, Wang B, Selleck EM, Sommer MOA, Dantas G (2012) The shared antibiotic resistome of soil bacteria and human pathogens. Science 337:1107–1111. doi:10.1126/science.1220761

Fritz J, Zuo Y (2007) Simultaneous determination of tetracycline, oxytetracycline, and 4-epitetracycline in milk by high-performance liquid chromatography. Food Chem 105:1297–1301. doi:10.1016/j.foodchem.2007.03.047

Furtula V, Stephenson GL, Olaveson KM, Chambers PA (2012) Effects of the veterinary pharmaceutical salinomycin and its formulation on the plant Brassica rapa. Arch Environ Contam Toxicol 63:513–522. doi:10.1007/s00244-012-9807-y

García-Galán MJ, Díaz-Cruz MS, Barceló D (2008) Identification and determination of metabolites and degradation products of sulfonamide antibiotics. Trends Anal Chem 27:1008–1022. doi:10.1016/j.trac.2008.10.001

García-Galán MJ, Díaz-Cruz MS, Barceló D (2009) Combining chemical analysis and ecotoxicity to determine environmental exposure and to assess risk from sulfonamides. Trends Anal Chem 28:804–819. doi:10.1016/j.trac.2009.04.006

García-Galán MJ, Garrido T, Fraile J, Ginebreda A, Díaz-Cruz MS, Barceló D (2010) Simultaneous occurrence of nitrates and sulfonamide antibiotics in two ground water bodies of Catalonia (Spain). J Hydrol 383:93–101. doi:10.1016/j.jhydrol.2009.06.042

García-Santiago X, Franco-Uría A, Omil F, Lema JM (2016) Risk assessment of persistent pharmaceuticals in biosolids: dealing with uncertainty. J Hazard Mater 302:72–81. doi:10.1016/j.jhazmat.2015.09.035

Gerke HH (2006) Preferential flow descriptions for structured soils. J Plant Nutr Soil Sci 169:382–400. doi:10.1002/jpln.200521955

Ghosh S, LaPara TM (2007) The effects of subtherapeutic antibiotic use in farm animals on the proliferation and persistence of antibiotic resistance among soil bacteria. ISME J 1:191–203. doi:10.1038/ismej.2007.31

Gillings MR, Stokes HW (2012) Are humans increasing bacterial evolvability? Trends Ecol Evol 27:346–352. doi:10.1016/j.tree.2012.02.006

Girardi C, Greve J, Lamshöft M, Fetzer I, Miltner A, Schäffer A, Kästner M (2011) Biodegradation of ciprofloxacin in water and soil and its effects on the microbial communities. J Hazard Mater 198:22–30. doi:10.1016/j.jhazmat.2011.10.004

Glasener KM (2002) Why is soil important? Soil Sci Soc Am Sci. https://soils.org/files/science-policy/sssa-marketing-2013.pdf

Golet, E.M., Strehler, A., Alder A.C., Giger, W., (2002) Determination of Fluoroquinolone Antibacterial Agents in Sewage Sludge and Sludge-Treated Soil Using Accelerated Solvent Extraction Followed by Solid-Phase Extraction, Anal. Chem. 74, 5455–5462, 10.1021/ac025762m

Götz A, Smalla K (1997) Manure enhances plasmid mobilization and survival of Pseudomonas putida introduced into field soil. Appl Environ Microbiol 63:1980–1986 PMCID: PMC1389163

Grassi M, Rizzo L, Farina A (2013) Endocrine disruptors compounds, pharmaceuticals and personal care products in urban wastewater: implications for agricultural reuse and their removal by adsorption process. Environ Sci Pollut Res 20:3616–3628. doi:10.1007/s11356-013-1636-7

Grathwohl P, Susset B (2009) Comparison of percolation to batch and sequential leaching tests: theory and data. Waste Manag 29:2681–2688. doi:10.1016/j.wasman.2009.05.016

Grathwohl P, van der Sloot HA (2007) Groundwater risk assessment at contaminated sites (GRACOS): test methods and modeling approaches. In: Quevauviller P (ed) Groundwater science and policy. RSC, Cambridge

Grave K, Torren-Edo J, Muller A, Greko C, Moulin G, Mackay D, Fuchs K, Laurier L, Iliev D, Pokludová L, Genakritis M, Jacobsen E, Kurvits K, Kivilahti-Mäntylä K, Wallmann J, Kovács J, Lenharðsson JM, Beechinor JG, Perrella A, Mičule G, Zymantaite U, Meijering A, Prokopiak D, Ponte MH, Svetlin A, Hederová J, Madero CM, Girma K, Eckford S (2014) Variations in the sales and sales patterns of veterinary antimicrobial agents in 25 European countries. J Antimicrob Chemother. dku106. doi: 10.1093/jac/dku106

Haller MY, Müller SR, McArdell CS, Alder AC, Suter MJF (2002) Quantification of veterinary antibiotics (sulfonamides and trimethoprim) in animal manure by liquid chromatography-mass spectrometry. J Chromatogr A 952:111–120. doi:10.1016/S0021-9673(02)00083-3

Halling-Sørensen B (2001) Inhibition of aerobic growth and nitrification of bacteria in sewage sludge by antibacterial agents. Arch Environ Contam Toxicol 40:451–460. doi:10.1007/s002440010197

Halling-Sørensen B, Lykkeberg A, Ingerslev F, Blackwell P, Tjørnelund J (2003a) Characterisation of the abiotic degradation pathways of oxytetracyclines in soil interstitial water using LC-MS-MS. Chemosphere 50:1331–1342. doi:10.1016/S0045-6535(02)00766-X

Halling-Sørensen B, Sengeløv G, Ingerslev F, Jensen LB (2003b) Reduced antimicrobial potencies of oxytetracycline, tylosin, sulfadiazin, streptomycin, ciprofloxacin, and olaquindox due to environmental processes. Arch Environ Contam Toxicol 44:7–16. doi:10.1007/s00244-002-1234-z

Halling-Sørensen B, Jacobsen A-M, Jensen J, Sengeløv G, Vaclavik E, Ingerslev F (2005) Dissipation and effects of chlortetracycline and tylosin in two agricultural soils: a field-scale study in southern Denmark. Environ Toxicol Chem 24:802–810. doi:10.1897/03-576.1

Hammesfahr U, Heuer H, Manzke B, Smalla K, Thiele-Bruhn S (2008) Impact of the antibiotic sulfadiazine and pig manure on the microbial community structure in agricultural soils. Soil Biol Biochem 40:1583–1591. doi:10.1016/j.soilbio.2008.01.010

Hammesfahr U, Kotzerke A, Lamshöft M, Wilke B-M, Kandeler E, Thiele-Bruhn S (2011) Effects of sulfadiazine-contaminated fresh and stored manure on a soil microbial community. Eur J Soil Biol 47:61–68. doi:10.1016/j.ejsobi.2010.10.004

Hamscher G, Sczesny S, Höper H, Nau H (2002) Determination of persistent tetracycline residues in soil fertilized with liquid manure by high-performance liquid chromatography with electrospray ionization tandem mass spectrometry. Anal Chem 74:1509–1518. doi:10.1021/ac015588m

Hawker DW, Cropp R, Boonsaner M (2013) Uptake of zwitterionic antibiotics by rice (Oryza sativa L.) in contaminated soil. J Hazard Mater 263:458–466. doi:10.1016/j.jhazmat.2013.09.066

He X, Wang Z, Nie X, Yang Y, Pan D, Leung AOW, Cheng Z, Yang Y, Li K, Chen K (2012) Residues of fluoroquinolones in marine aquaculture environment of the Pearl River Delta, South China. Environ Geochem Health 34:323–335. doi:10.1007/s10653-011-9420-4

Herklotz PA, Gurung P, Vanden Heuvel B, Kinney CA (2010) Uptake of human pharmaceuticals by plants grown under hydroponic conditions. Chemosphere 78:1416–1421. doi:10.1016/j.chemosphere.2009.12.048

Heuer H, Smalla K (2012) Plasmids foster diversification and adaptation of bacterial populations in soil. FEMS Microbiol Rev 36:1083–1104. doi:10.1111/j.1574-6976.2012.00337.x

Heuer H, Focks A, Lamshöft M, Smalla K, Matthies M, Spiteller M (2008) Fate of sulfadiazine administered to pigs and its quantitative effect on the dynamics of bacterial resistance genes in manure and manured soil. Soil Biol Biochem 40:1892–1900. doi:10.1016/j.soilbio.2008.03.014

Heuer H, Schmitt H, Smalla K (2011) Antibiotic resistance gene spread due to manure application on agricultural fields. Curr Opin Microbiol 14:236–243. doi:10.1016/j.mib.2011.04.009

Hiba A, Carine A, Haifa AR, Ryszard L, Farouk J (2016) Monitoring of twenty-two sulfonamides in edible tissues: investigation of new metabolites and their potential toxicity. Food Chem 192:212–227. doi:10.1016/j.foodchem.2015.06.093

Ho YB, Zakaria MP, Latif PA, Saari N (2012) Simultaneous determination of veterinary antibiotics and hormone in broiler manure, soil and manure compost by liquid chromatography-tandem mass spectrometry. J Chromatogr A 1262:160–168. doi:10.1016/j.chroma.2012.09.024

Ho YB, Zakaria MP, Latif PA, Saari N (2013) Degradation of veterinary antibiotics and hormone during broiler manure composting. Bioresour Technol 131:476–484. doi:10.1016/j.biortech.2012.12.194

Hoagland RE (1996) Herbicidal properties of the antibiotic monensin. J Sci Food Agric 70:373–379. doi:10.1002/(SICI)1097-0010(199603)70:3<373::AID-JSFA516>3.0.CO;2-P

Hoagland RE, Zablotowicz RM, Hall JC (2000) Pesticide metabolism in plants and microorganisms: an overview. Pestic Biotransformation Plants Microorg 777:2–27. doi:10.1021/bk-2001-0777.ch001

Holling CS, Bailey JL, Vanden Heuvel B, Kinney CA (2012) Uptake of human pharmaceuticals and personal care products by cabbage (Brassica campestris) from fortified and biosolids-amended soils. J Environ Monit 14:3029. doi:10.1039/c2em30456b

Homem V, Santos L (2011) Degradation and removal methods of antibiotics from aqueous matrices—a review. J Environ Manage 92:2304–2347. doi:10.1016/j.jenvman.2011.05.023

Hong PY, Yannarell AC, Dai Q, Ekizoglu M, Mackie RI (2013) Monitoring the perturbation of soil and groundwater microbial communities due to pig production activities. Appl Environ Microbiol 79:2620–2629. doi:10.1128/AEM.03760-12

Horvat AJM, Babić S, Pavlović DM, Ašperger D, Pelko S, Kaštelan-Macan M, Petrović M, Mance AD (2012) Analysis, occurrence and fate of anthelmintics and their transformation products in the environment. TrAC Trends Anal. Chem. 31:61–84. doi:10.1016/j.trac.2011.06.023

Hu X, Luo Y, Zhou Q (2010a) Simultaneous analysis of selected typical antibiotics in manure by microwave-assisted extraction and LC–MS n. Chromatographia 71:217–223. doi:10.1365/s10337-009-1438-8

Hu X, Zhou Q, Luo Y (2010b) Occurrence and source analysis of typical veterinary antibiotics in manure, soil, vegetables and groundwater from organic vegetable bases, northern China. Environ Pollut 158:2992–2998. doi:10.1016/j.envpol.2010.05.023

Hu W, Ma L, Guo C, Sha J, Zhu X (2012) Simultaneous extraction and determination of fluoroquinolones, tetracyclines and sulfonamides antibiotics in soils using optimised solid phase extraction chromatography-tandem mass spectrometry. Int J Environ Anal Chem 37–41. doi: 10.1080/03067319.2010.520122

Huang X, Liu C, Li K, Liu F, Liao D, Liu L, Zhu G, Liao J (2013) Occurrence and distribution of veterinary antibiotics and tetracycline resistance genes in farmland soils around swine feedlots in Fujian Province, China. Environ Sci Pollut Res 20:9066–9074. doi:10.1007/s11356-013-1905-5

Huber C, Bartha B, Schroder P (2012) Metabolism of diclofenac in plants—hydroxylation is followed by glucose conjugation. J Hazard Mater 243:250–256. doi:10.1016/j.jhazmat.2012.10.023

Hund-Rinke K, Simon M, Lukow T (2004) Effects of tetracycline on the soil microflora: function, diversity, resistance. J Soil Sediment 4:11–16. doi:10.1007/BF02990823

Huysman F, Van Renterghem B, Verstraete W (1993) Antibiotic resistant sulphite-reducing clostridia in soil and groundwater as indicator of manuring practices. Water Air Soil Pollut 69:243–255. doi:10.1007/BF00478161

Ingerslev F, Halling-Sørensen B (2001) Biodegradability of metronidazole, olaquindox, and tylosin and formation of tylosin degradation products in aerobic soil—manure slurries. Ecotoxicol Environ Saf 48:311–320. doi:10.1006/eesa.2000.2026

Ingerslev F, Toräng L, Loke ML, Halling-Sorensen B, Nyholm N (2001) Primary biodegradation of veterinary antibiotics in aerobic and anaerobic surface water simulation systems. Chemosphere 44:865–872. doi:10.1016/S0045-6535(00)00479-3

Inoue J, Chamberlain K, Bromilow RH (1998) Physicochemical factors affecting the uptake by roots and translocation to shoots of amine bases in Barley. Pestic Sci 54:8–21. doi:10.1002/(SICI)1096-9063(199809)54:1<8::AID-PS793>3.0.CO;2-E

Institute of Medicine (1989) Human health risks with the subtherapeutic use of penicillin or tetracyclines in animal feed. National Academy Press, Washington, DC

Jacobsen AM, Halling-Sørensen B, Ingerslev F, Hansen SH (2004) Simultaneous extraction of tetracycline, macrolide and sulfonamide antibiotics from agricultural soils using pressurised liquid extraction, followed by solid-phase extraction and liquid chromatography-tandem mass spectrometry. J Chromatogr A 1038:157–170. doi:10.1016/j.chroma.2004.03.034

Jechalke S, Heuer H, Siemens J, Amelung W, Smalla K (2014) Fate and effects of veterinary antibiotics in soil. Trends Microbiol 22:536–545. doi:10.1016/j.tim.2014.05.005

Jia A, Xiao Y, Hu J, Asami M, Kunikane S (2009) Simultaneous determination of tetracyclines and their degradation products in environmental waters by liquid chromatography-electrospray tandem mass spectrometry. J Chromatogr A 1216:4655–4662. doi:10.1016/j.chroma.2009.03.073

Jiaa S, Hea X, Bub Y, Shia P, Miaoa Y, Zhoub H, Shanb Z, Zhang XX (2014) Environmental fate of tetracycline resistance genes originating from swine feedlots in river water. J Environ Sci Health B 49:624–631. doi:10.1080/03601234.2014.911594

Jjemba PK (2002) The potential impact of veterinary and human therapeutic agents in manure and biosolids on plants grown on arable land: a review. Agric Ecosyst Environ 93:267–278. doi:10.1016/S0167-8809(01)00350-4

Johnsen PJ, Townsend JP, Bøhn T, Simonsen GS, Sundsfjord A, Nielsen KM (2009) Factors affecting the reversal of antimicrobial-drug resistance. Lancet Infect Dis 9(6):357–364. doi:10.1016/S1473-3099(09)70105-7

Jones-Lepp TL, Sanchez CA, Moy T, Kazemi R (2010) Method development and application to determine potential plant uptake of antibiotics and other drugs in irrigated crop production systems. J Agric Food Chem 58:11568–11573. doi:10.1021/jf1028152

Joy SR, Li X, Snow DD, Gilley JE, Woodbury B, Bartelt-Hunt SL (2014) Fate of antimicrobials and antimicrobial resistance genes in simulated swine manure storage. Sci Total Environ 481:69–74. doi:10.1016/j.scitotenv.2014.02.027

K'oreje KO, Vergeynst L, Ombaka D, De Wispelaere P, Okoth M (2016) Occurrence patterns of pharmaceutical residues in wastewater, surface water and groundwater of Nairobi and Kisumu city, Kenya. Chemosphere 149:238–244. doi:10.1016/j.chemosphere.2016.01.095

Kang DH, Gupta S, Rosen C, Fritz V, Singh A, Chander Y, Murray H, Rohwer C (2013) Antibiotic uptake by vegetable crops from manure-applied soils. J Agric Food Chem 61:9992–10001. doi:10.1021/jf404045m

Karci A, Balcioğlu IA (2009) Investigation of the tetracycline, sulfonamide, and fluoroquinolone antimicrobial compounds in animal manure and agricultural soils in Turkey. Sci Total Environ 407:4652–4664. doi:10.1016/j.scitotenv.2009.04.047

Kay P, Blackwell PA, Boxall ABA (2004) Fate of veterinary antibiotics in a macroporous tile drained clay soil. Environ Toxicol Chem 23, 1136–1144. doi: 10.1897/03-374

Kay P, Blackwell PA, Boxall ABA (2005a) Column studies to investigate the fate of veterinary antibiotics in clay soils following slurry application to agricultural land. Chemosphere 60:497–507. doi:10.1016/j.chemosphere.2005.01.028

Kay P, Blackwell PA, Boxall ABA (2005b) A lysimeter experiment to investigate the leaching of veterinary antibiotics through a clay soil and comparison with field data. Environ Pollut 134:333–341. doi:10.1016/j.envpol.2004.07.021

Kelessidis A, Stasinakis AS (2012) Comparative study of the methods used for treatment and final disposal of sewage sludge in European countries. Waste Manag 32:1186–1195. doi:10.1016/j.wasman.2012.01.012

Kemper N (2008) Veterinary antibiotics in the aquatic and terrestrial environment. Ecol Indic 8:1–13. doi:10.1016/j.ecolind.2007.06.002

Kemper N, Faber H, Skutlarek D, Krieter J (2008) Analysis of antibiotic residues in liquid manure and leachate of dairy farms in Northern Germany. Agric Water Manag 95:1288–1292 doi:10.1016/j.agwat.2008.05.008

Kim SC, Yang JE, Ok YS, Carlson K (2010) Dissolved and colloidal fraction transport of antibiotics in soil under biotic and abiotic conditions. Water Qual Res J Canada 45:275–285

Knapp CW, Dolfing J, Ehlert PAI, Graham DW (2010) Evidence of increasing antibiotic resistance gene abundances in archived soils since 1940. Environ Sci Technol 44:580–587. doi:10.1021/es901221x

Kołodziejska M, Maszkowska J, Białk-Bielińska A, Steudte S, Kumirska J, Stepnowski P, Stolte S (2013) Aquatic toxicity of four veterinary drugs commonly applied in fish farming and animal husbandry. Chemosphere 92:1253–1259. doi:10.1016/j.chemosphere.2013.04.057

Kolz AC, Moorman TB, Ong SK, Scoggin KD, Douglass EA (2005) Degradation and metabolite production of tylosin in anaerobic and aerobic swine-manure lagoons. Water Environ Res 77:49–56. doi:10.2175/106143005X41618

Kong WD, Zhu YG, Liang YC, Zhang J, Smith FA, Yang M (2007) Uptake of oxytetracycline and its phytotoxicity to alfalfa (Medicago sativa L.). Environ Pollut 147:187–193. doi:10.1016/j.envpol.2006.08.016

Kools SAE, Moltmann JF, Knacker T (2008) Estimating the use of veterinary medicines in the European union. Regul Toxicol Pharmacol 50:59–65. doi:10.1016/j.yrtph.2007.06.003

Kotzerke A, Sharma S, Schauss K, Heuer H, Thiele-Bruhn S, Smalla K, Wilke BM, Schloter M (2008) Alterations in soil microbial activity and N-transformation processes due to sulfadiazine loads in pig-manure. Environ Pollut 153:315–322. doi:10.1016/j.envpol.2007.08.020

Kuchta SL, Cessna AJ (2009) Lincomycin and spectinomycin concentrations in liquid swine manure and their persistence during simulated manure storage. Arch Environ Contam Toxicol 57:1–10. doi:10.1007/s00244-008-9229-z

Kumar K, Gupta SC, Baidoo SK, Chander Y, Rosen CJ (2005) Antibiotic uptake by plants from soil fertilized with animal manure. J Environ Qual 34:2082–2085. doi:10.2134/jeq2005.0026

Kummerer K (2003) Significance of antibiotics in the environment. J Antimicrob Chemother 52:5–7. doi:10.1093/jac/dkg293

Kurwadkar ST, Adams CD, Meyer MT, Kolpin DW (2007) Effects of sorbate speciation on sorption of selected sulfonamides in three loamy soils. J Agric Food Chem 55:1370–1376. doi:10.1021/jf060612o

Kurwadkar ST, Adams CD, Meyer MT, Kolpin DW (2011) Comparative mobility of sulfonamides and bromide tracer in three soils. J Environ Manage 92:1874–1881. doi:10.1016/j.jenvman.2011.03.018

Kuster M, José López de Alda M, Barceló D (2004) Analysis and distribution of estrogens and progestogens in sewage sludge, soils and sediments. TrAC Trends Anal. Chem. 23:790–798. doi:10.1016/j.trac.2004.08.007

Küster A, Bachmann J, Brandt U, Ebert I, Hickmann S, Klein-Goedicke J, Maack G, Schmitz S, Thumm E, Rechenberg B (2009) Regulatory demands on data quality for the environmental risk assessment of pharmaceuticals. Regul Toxicol Pharmacol 55:276–280. doi:10.1016/j.yrtph.2009.07.005

Kwon SI, Owens G, Ok YS, Lee DB, Jeon W-T, Kim JG, Kim K-R (2011) Applicability of the Charm II system for monitoring antibiotic residues in manure-based composts. Waste Manag 31:39–44. doi:10.1016/j.wasman.2010.08.018

Lapworth DJ, Baran N, Stuart ME, Ward RS (2012) Emerging organic contaminants in groundwater: a review of sources, fate and occurrence. Environ Pollut 163:287–303. doi:10.1016/j.envpol.2011.12.034

Le-Minh N, Khan SJ, Drewes JE, Stuetz RM (2010) Fate of antibiotics during municipal water recycling treatment processes. Water Res 44:4295–4323. doi:10.1016/j.watres.2010.06.020

Li WC (2014) Occurrence, sources, and fate of pharmaceuticals in aquatic environment and soil. Environ Pollut 187:193–201. doi:10.1016/j.envpol.2014.01.015

Li YW, Wu XL, Mo CH, Tai YP, Huang XP, Xiang L (2011) Investigation of sulfonamide, tetracycline, and quinolone antibiotics in vegetable farmland soil in the Pearl River Delta area, southern China. J Agric Food Chem 59:7268–7276. doi:10.1021/jf1047578

Liguoro M, Cibin V, Capolongo F, Halling-Sørensen B, Montesissa C (2003) Use of oxytetracycline and tylosin in intensive calf farming: evaluation of transfer to manure and soil Marco. Chemosphere 68:671–676. doi:10.1016/j.chemosphere.2007.02.009

Lillenberg M, Yurchenko S, Kipper K, Herodes K, Pihl V, Sepp K, Lõhmus R, Nei L (2009) Simultaneous determination of fluoroquinolones, sulfonamides and tetracyclines in sewage sludge by pressurized liquid extraction and liquid chromatography electrospray ionization-mass spectrometry. J Chromatogr A 1216:5949–5954. doi:10.1016/j.chroma.2009.06.029

Lillicrap A, Macken A, Thomas KV (2015) Recommendations for the inclusion of targeted testing to improve the regulatory environmental risk assessment of veterinary medicines used in aquaculture. Environ Int 85:1–4. doi:10.1016/j.envint.2015.07.019

Lindsey ME, Meyer M, Thurman EM (2001) Analysis of trace levels of sulfonamide and tetracycline antimicrobials in groundwater and surface water using solid-phase extraction and liquid chromatography/mass spectrometry. Anal Chem 73:4640–4646. doi:10.1021/ac010514w

Liu F, Ying GG, Tao R, Zhao JL, Yang JF, Zhao LF (2009) Effects of six selected antibiotics on plant growth and soil microbial and enzymatic activities. Environ Pollut 157:1636–1542. doi:10.1016/j.envpol.2008.12.021

Liu L, Liu YH, Liu CX, Wang Z, Dong J, Zhu GF, Huang X (2013a) Potential effect and accumulation of veterinary antibiotics in Phragmites australis under hydroponic conditions. Ecol Eng 53:138–143. doi:10.1016/j.ecoleng.2012.12.033

Liu D, Lung W-S, Colosi LM (2013b) Effects of sorption kinetics on the fate and transport of pharmaceuticals in estuaries. Chemosphere 92:1001–1009. doi:10.1016/j.chemosphere.2013.03.029

Liu FJ, Li SX, Zheng FY, Huang XG, Zuo YG, Tu TX, Wu XQ (2014) Risk assessment of nitrate and oxytetracycline addition on coastal ecosystem functions. Aquat Toxicol 146:76–81. doi:10.1016/j.aquatox.2013

Loke ML, Ingerslev F, Halling-Sørensen B, Tjørnelund J (2000) Stability of Tylosin A in manure containing test systems determined by high performance liquid chromatography. Chemosphere 40:759–765. doi:10.1016/S0045-6535(99)00450-6

Lopez Meza S, Garrabrants AC, Van der Sloot H, Kosson DS (2008) Comparison of the release of constituents from granular materials under batch and column testing. Waste Manag 28:1853–1867. doi:10.1016/j.wasman.2007.11.009

López-Serna R, Jurado A, Vázquez-Suñé E, Carrera J, Petrović M, Barceló D (2013) Occurrence of 95 pharmaceuticals and transformation products in urban groundwaters underlying the metropolis of Barcelona. Spain Environ Pollut 174:305–315. doi:10.1016/j.envpol.2012.11.022

MacHerius A, Eggen T, Lorenz W, Moeder M, Ondruschka J, Reemtsma T (2012) Metabolization of the bacteriostatic agent triclosan in edible plants and its consequences for plant uptake assessment. Environ Sci Technol 46:10797–10804. doi:10.1021/es3028378

Maia AS, Ribeiro AR, Amorim CL, Barreiro JC, Cass QB, Castro PML, Tiritan ME (2014) Degradation of fluoroquinolone antibiotics and identification of metabolites/transformation products by liquid chromatography-tandem mass spectrometry. J Chromatogr A 1333:87–98. doi:10.1016/j.chroma.2014.01.069

Marsoni M, De Mattia F, Labra M, Bruno A, Bracale M, Vannini C (2014) Uptake and effects of a mixture of widely used therapeutic drugs in Eruca sativa L. and Zea mays L. plants. Ecotoxicol Environ Saf 108:52–57. doi:10.1016/j.ecoenv.2014.05.029

Martínez JL (2008) Antibiotics and antibiotic resistance genes in natural environments. Science 321:365–367. doi:10.1126/science.1159483

Martínez-Carballo E, González-Barreiro C, Scharf S, Gans O (2007) Environmental monitoring study of selected veterinary antibiotics in animal manure and soils in Austria. Environ Pollut 148:570–579. doi:10.1016/j.envpol.2006.11.035

Masse DI, Saady NMC, Gilbert Y (2014) Potential of biological processes to eliminate antibiotics in livestock manure: an overview. Animals 4:146–163. doi:10.3390/ani4020146

Maszkowska J, Kołodziejska M, Białk-Bielińska A, Mrozik W, Kumirska J, Stepnowski P, Palavinskas R, Krüger O, Kalbe U (2013) Column and batch tests of sulfonamide leaching from different types of soil. J Hazard Mater 260:468–474. doi:10.1016/j.jhazmat.2013.05.053

Maszkowska J, Wagil M, Mioduszewska K, Kumirska J, Stepnowski P, Białk-Bielińska A (2014) Thermodynamic studies for adsorption of ionizable pharmaceuticals onto soil. Chemosphere 111:568–574. doi:10.1016/j.chemosphere.2014.05.005

Matamoros V, Nguyen LX, Arias CA, Salvado V, Brix H (2012) Evaluation of aquatic plants for removing polar microcontaminants: a microcosm experiment. Chemosphere 88:1257–1264. doi:10.1016/j.chemosphere.2012.04.004

Metcalfe CD, Alder AC, Halling-Sørensen B, Krogh K, Fenner K, Larsbo M, Straub JO, Ternes TA, Topp E, Lapen DR, Boxall ABA (2008) Exposure assessment methods for veterinary and human-use medicines in the environment: PEC vs. MEC comparisons. Chapter 11. In:

Pharmaceuticals in the environment. Sources, fate, effects and risks. pp 147–171. doi: 10.1007/978-3-540-74664-5_11

Michelini L, La Rocca N, Rascio N, Ghisi R (2013) Structural and functional alterations induced by two sulfonamide antibiotics on barley plants. Plant Physiol Biochem 67:55–62. doi:10.1016/j.plaphy.2013.02.027

Migliore L, Cozzolino S, Fiori M (2003) Phytotoxicity to and uptake of enrofloxacin in crop plants. Chemosphere 52:1233–1244. doi:10.1016/S0045-6535(03)00272-8

Miller EL, Nason SL, Karthikeyan K, Pedersen JA (2015) Root uptake of pharmaceuticals and personal care product ingredients. Environ Sci Technol 50:acs.est.5b01546. doi:10.1021/acs.est.5b01546

Ministry of Agriculture and Rural Development (2013) Antimicrobial veterinary medicinal products in 2012 in Poland. Warsaw, Poland

Mitchell SM, Ullman JL, Teel AL, Watts RJ (2014) pH and temperature effects on the hydrolysis of three β-lactam antibiotics: ampicillin, cefalotin and cefoxitin. Sci Total Environ 466–467:547–555. doi:10.1016/j.scitotenv.2013.06.027

Montesdeoca-Esponda S, Sosa-Ferrera Z, Santana-Rodríguez JJ (2012) Combination of microwave-assisted micellar extraction with liquid chromatography tandem mass spectrometry for the determination of fluoroquinolone antibiotics in coastal marine sediments and sewage sludges samples. Biomed Chromatogr 26:33–40. doi:10.1002/bmc.1621

Montforts MHMM (1999) Environmental risk assessment for veterinary medicinal products part 1. Non-immunological drug substances. Second update RIVM report 320202001/2003. Bilthoven

Moore MT, Kröger R (2010) Effect of three insecticides and two herbicides on rice (Oryza sativa) seedling germination and growth. Arch Environ Contam Toxicol 59:574–581. doi:10.1007/s00244-010-9519-0

Moral R, Paredes C, Bustamante MA, Marhuenda-Egea F, Bernal MP (2009) Utilisation of manure composts by high-value crops: safety and environmental challenges. Bioresour Technol 100:5454–5460. doi:10.1016/j.biortech.2008.12.007

Motarjemi Y, Moy G, Tood E (2014) Encyclopedia of food safety, vol 3. Elsevier, San Diego

Muller AK, Westergaard K, Christensen S, Sorensen SJ (2002) The diversity and function of soil microbial communities exposed to different disturbances. Microb Ecol 44:49–58. doi:10.1007/s00248-001-0042-8

Nebot C, Guarddon M, Seco F, Iglesias A, Miranda JM, Franco CM, Cepeda A (2014) Monitoring the presence of residues of tetracyclines in baby food samples by HPLC-MS/MS. Food Control 46:495–501. doi:10.1016/j.foodcont.2014.05.042

Nesme J, Simonet P (2015) The soil resistome: a critical review on antibiotic resistance origins, ecology and dissemination potential in telluric bacteria. Environ Microbiol 17:913–930. doi:10.1111/1462-2920.12631

Nesme J, Cecillon S, Delmont TO, Monier JM, Vogel TM, Simonet P (2014) Large-scale metagenomic-based study of antibiotic resistance in the environment. Curr Biol 24:1096–1100. doi:10.1016/j.cub.2014.03.036

O'Connor S, Aga DS (2007) Analysis of tetracycline antibiotics in soil: advances in extraction, clean-up, and quantification. Trends Anal Chem 26:456–465. doi:10.1016/j.trac.2007.02.007

Ollivier J, Kleineidam K, Kotzerke A, Kindler R, Wilke B, Schloter M (2010) Effect of sulfadiazine-contaminated pig manure on the abundances of genes and transcripts involved in nitrogen transformation in the root-rhizosphere complexes of maize and clover. Appl Environ Microbiol 76:7903–7909. doi:10.1128/AEM.01252-10

Oppel J, Broll G, Loffker D, Meller M, Rombke J, Ternes T, Löffler D, Meller M, Römbke J, Ternes T (2004) Leaching behaviour of pharmaceuticals in soil-testing-systems: a part of an environmental risk assessment for groundwater protection. Sci Total Environ 328:265–273. doi:10.1016/j.scitotenv.2004.02.004

Opris O, Copaciu F, Loredana Soran M, Ristoiu D, Niinemets U, Copolovici L (2013) Influence of nine antibiotics on key secondary metabolites and physiological characteristics in Triticum aestivum: leaf volatiles as a promising new tool to assess toxicity. Ecotoxicol Environ Saf 87:70–79. doi:10.1016/j.ecoenv.2012.09.019

Ostermann A, Siemens J, Welp G, Xue Q, Lin X, Liu X, Amelung W (2013) Leaching of veterinary antibiotics in calcareous Chinese croplands. Chemosphere 91:928–934. doi:10.1016/j.chemosphere.2013.01.110

Pan M, Chu LM (2016a) Adsorption and degradation of five selected antibiotics in agricultural soil. Sci Total Environ 545–546:48–56. doi:10.1016/j.scitotenv.2015.12.040

Pan M, Chu LM (2016b) Phytotoxicity of veterinary antibiotics to seed germination and root elongation of crops. Ecotoxicol Environ Saf 126:228–237. doi:10.1016/j.ecoenv.2015.12.027

Pan X, Qiang Z, Ben W, Chen M (2011) Residual veterinary antibiotics in swine manure from concentrated animal feeding operations in Shandong Province, China. Chemosphere 84:695–700. doi:10.1016/j.chemosphere.2011.03.022

Pan M, Wong CKC, Chu LM (2014) Distribution of antibiotics in wastewater-irrigated soils and their accumulation in vegetable crops in the Pearl River Delta, southern China. J Agric Food Chem 62:11062–11069. doi:10.1021/jf503850v

Peng S, Wang Y, Zhou B, Lin X (2015) Long-term application of fresh and composted manure increase tetracycline resistance in the arable soil of eastern China. Sci Total Environ 506–507:279–286. doi:10.1016/j.scitotenv.2014.11.010

Picó Y, Andreu V (2007) Fluoroquinolones in soil-risks and challenges. Anal Bioanal Chem 387:1287–1299. doi:10.1007/s00216-006-0843-1

Pino MR, Val J, Mainar AM, Zuriaga E, Espanol C, Langa E (2015) Acute toxicological effects on the earthworm Eisenia fetida of 18 common pharmaceuticals in artificial soil. Sci Total Environ 518–519:225–237. doi:10.1016/j.scitotenv.2015.02.080

Podlipná R, Skálová L, Seidlová H, Szotáková B, Kubíček V, Stuchlíková L, Jirásko R, Vaněk T, Vokřál I (2013) Biotransformation of benzimidazole anthelmintics in reed (Phragmites australis) as a potential tool for their detoxification in environment. Bioresour Technol 144:216–224. doi:10.1016/j.biortech.2013.06.105

Pomati F, Netting AG, Calamari D, Neilan BA (2004) Effects of erythromycin, tetracycline and ibuprofen on the growth of Synechocystis sp. and Lemna minor. Aquat Toxicol 67:387–396. doi:10.1016/j.aquatox.2004.02.001

Popova IE, Bair DA, Tate KW, Sanjai P (2014) Sorption, leaching, and surface runoff of beef cattle veterinary pharmaceuticals under simulated irrigated pasture conditions. J Environ Qual 42:1167–1175. doi:10.2134/jeq2013.01.0012

Poskrobko B, Poskrobko T, Skiba K (2007) Zagrożenia i Ochrona Gleb. PWN, Warszawa

Prosser RS, Sibley PK (2015) Human health risk assessment of pharmaceuticals and personal care products in plant tissue due to biosolids and manure amendments, and wastewater irrigation. Environ Int 84:209–212. doi:10.1016/j.envint.2015.07.007

Pruden A, Arabi M, Storteboom HN (2012) Correlation between upstream human activities and riverine antibiotic resistance genes. Environ Sci Technol 46:11541–11549. doi:10.1021/es302657r

Pruneda A (1950) The medical impact of antimicrobial use in food animals. Report of a WHO meeting. Medicina (B. Aires) 30:175–178. doi: 10.5860/CHOICE.41-4081

Qiao M, Chen W, Su J, Zhang B, Zhang C (2012) Fate of tetracyclines in swine manure of three selected swine farms in China. J Environ Sci 24:1047–1052. doi:10.1016/S1001-0742(11)60890-5

Rabølle M, Spliid NH (2000) Sorption and mobility of metronidazole, olaquindox, oxytetracycline and tylosin in soil. Chemosphere 40:715–722. doi:10.1016/S0045-6535(99)00442-7

Raich-Montiu J, Beltrán JL, Prat MD, Granados M (2010) Studies on the extraction of sulfonamides from agricultural soils. Anal Bioanal Chem 397:807–814. doi:10.1007/s00216-010-3580-4

Raich-Montiu J, Prat MD, Granados M (2011) Extraction and analysis of avermectines in agricultural soils by microwave assisted extraction and ultra high performance liquid chromatography coupled to tandem mass spectrometry. Anal Chim Acta 697:32–37. doi:10.1016/j.aca.2011.04.021

Ramaswamy J, Prasher SO, Patel RM, Hussain SA, Barrington SF (2010a) The effect of composting on the degradation of a veterinary pharmaceutical. Bioresour Technol 101:2294–2299. doi:10.1016/j.biortech.2009.10.089

Ramaswamy J, Prasher SO, Patel RM, Hussain SA, Barrington SF (2010b) The effect of composting on the degradation of a veterinary pharmaceutical. Bioresour Technol 101:2294–2299. doi:10.1016/j.biortech.2009.10.089

Rehman MSU, Rashid N, Ashfaq M, Saif A, Ahmad N, Han J-I (2013) Global risk of pharmaceutical contamination from highly populated developing countries. Chemosphere 138:1045–1055. doi:10.1016/j.chemosphere.2013.02.036

Reichel R, Rosendahl I, Peeters ETHM, Focks A, Groeneweg J, Bierl R, Schlichting A, Amelung W, Thiele-Bruhn S (2013) Effects of slurry from sulfadiazine-(SDZ) and difloxacin-(DIF) medicated pigs on the structural diversity of microorganisms in bulk and rhizosphere soil. Soil Biol Biochem 62:82–91. doi:10.1016/j.soilbio.2013.03.007

Rentsch D, Schmidt S, Tegeder M (2007) Transporters for uptake and allocation of organic nitrogen compounds in plants. FEBS Lett 581:2281–2289. doi:10.1016/j.febslet.2007.04.013

Robinson AA, Belden JB, Lydy MJ (2005) Toxicity of fluoroquinolone antibiotics to aquatic organisms. Environ Toxicol Chem 24:423–430. doi:10.1897/04-210R.1

Roccaro P, Vagliasindi FGA (2014) Risk assessment of the use of biosolids containing emerging organic contaminants in agriculture. Chem Eng Trans 37:817–822. doi:10.3303/CET1437137

Rodriguez-Ruiz A, Etxebarria J, Boatti L, Marigomez I (2015) Scenario-targeted toxicity assessment through multiple endpoint bioassays in a soil posing unacceptable environmental risk according to regulatory screening values. Environ Sci Pollut Res 22:13344–13361. doi:10.1007/s11356-015-4564-x

Sabourin L, Duenk P, Bonte-Gelok S, Payne M, Lapen DR, Topp E (2012) Uptake of pharmaceuticals, hormones and parabens into vegetables grown in soil fertilized with municipal biosolids. Sci Total Environ 431:233–236. doi:10.1016/j.scitotenv.2012.05.017

Sarmah AK, Meyer MT, Boxall ABA (2006) A global perspective on the use, sales, exposure pathways, occurrence, fate and effects of veterinary antibiotics (VAs) in the environment. Chemosphere 65:725–759. doi:10.1016/j.chemosphere.2006.03.026

Sassman SA, Lee LS (2005) Sorption of three tetracyclines by several soils: assessing the role of pH and cation exchange. Environ Sci Technol 39:7452–7459. doi:10.1021/es0480217

Schauss K, Focks A, Leininger S, Kotzerke A, Heuer H, Thiele-Bruhn S, Sharma S, Wilke BM, Matthies M, Smalla K, Munch JC, Amelung W, Kaupenjohann M, Schloter M, Schleper C (2009) Dynamics and functional relevance of ammonia-oxidizing archaea in two agricultural soils. Environ Microbiol 11:446–456. doi:10.1111/j.1462-2920.2008.01783.x

Schlabach M, Grung M, Heimstad ES, Moe M, Schlabach M, Svenson A, Thomas K, Woldegiorgis A (2007) Human and veterinary pharmaceuticals, narcotics, and personal care products in the environment. Current state of knowledge and monitoring requirements. Oslo

Schlusener MP, Bester K (2006) Persistence of antibiotics such as macrolides, tiamulin and salinomycin in soil. Environ Pollut 143:565–571. doi:10.1016/j.envpol.2005.10.049

Schmitt H, Haapakangas H, Van Beelen P (2005) Effects of antibiotics on soil microorganisms: time and nutrients influence pollution-induced community tolerance. Soil Biol Biochem 37:1882–1892. doi:10.1016/j.soilbio.2005.02.022

Sengeløv G, Agersø Y, Halling-Sørensen B, Baloda SB, Andersen JS, Jensen LB (2003) Bacterial antibiotic resistance levels in Danish farmland as a result of treatment with pig manure slurry. Environ Int 28:587–595. doi:10.1016/S0160-4120(02)00084-3

Shelver WL, Hakk H, Larsen GL, DeSutter TM, Casey FXM (2010) Development of an ultra-high-pressure liquid chromatography-tandem mass spectrometry multi-residue sulfonamide method and its application to water, manure slurry, and soils from swine rearing facilities. J Chromatogr A 1217:1273–1282. doi:10.1016/j.chroma.2009.12.034

Shenker M, Harush D, Ben-Ari J, Chefetz B (2011) Uptake of carbamazepine by cucumber plants—a case study related to irrigation with reclaimed wastewater. Chemosphere 82:905–910. doi:10.1016/j.chemosphere.2010.10.052

Sittig S, Kasteel R, Groeneweg J, Hofmann D, Thiele B, Koppchen S, Vereecken H (2014) Dynamics of transformation of the veterinary antibiotic sulfadiazine in two soils. Chemosphere 95:470–477. doi:10.1016/j.chemosphere.2013.09.100

Smalla K, Heuer H, Gotz A, Niemeyer D, Krogerrecklenfort E, Tietze E (2000) Exogenous isolation of antibiotic resistance plasmids from piggery manure slurries reveals a high prevalence and diversity of IncQ-like plasmids. Appl Environ Microbiol 66:4854–4862. doi:10.1128/AEM.66.11.4854-4862.2000

Solliec M, Roy-Lachapelle A, Gasser M-O, Coté C, Généreux M, Sauvé S (2016) Fractionation and analysis of veterinary antibiotics and their related degradation products in agricultural soils and drainage waters following swine manure amendment. Sci Total Environ 543:524–535. doi:10.1016/j.scitotenv.2015.11.061

Song W, Guo M (2014) Applied manure and nutrient chemistry for sustainable agriculture and environment. In: He Z, Zhang H (eds) Applied manure and nutrient chemistry for sustainable agriculture and environment. pp 1–379. doi: 10.1007/978-94-017-8807-6

Speltini A, Sturini M, Maraschi F, Viti S, Sbarbada D, Profumo A (2015) Fluoroquinolone residues in compost by green enhanced microwave-assisted extraction followed by ultra performance liquid chromatography tandem mass spectrometry. J Chromatogr A 1410:44–50. doi:10.1016/j.chroma.2015.07.093

Spielmeyer A, Breier B, Groißmeier K, Hamscher G (2015) Elimination patterns of worldwide used sulfonamides and tetracyclines during anaerobic fermentation. Bioresour Technol 193:307–314. doi:10.1016/j.biortech.2015.06.081

Srinivasan P, Sarmah AK, Manley-Harris M (2014) Sorption of selected veterinary antibiotics onto dairy farming soils of contrasting nature. Sci Total Environ 472:695–703. doi:10.1016/j.scitotenv.2013.11.104

Steinfeld H (2004) The livestock revolution—a global veterinary mission. Vet Parasitol 125:19–41. doi:10.1016/j.vetpar.2004.05.003

Sterling TM (1994) Mechanisms of herbicide absorption across plant membranes and accumulation in plant cells in plant cells' and accumulation across plant membranes of herbicide absorption mechanisms move across plant membranes. Weed Sci 42:263–276

Storteboom HN, Kim SC, Doesken KC, Carlson KH, Davis JG, Pruden A (2007) Response of antibiotics and resistance genes to high-intensity and low-intensity manure management. J Environ Qual 36:1695. doi:10.2134/jeq2007.0006

Sui Q, Cao X, Lu S, Zhao W, Qiu Z, Yu G (2015) Occurrence, sources and fate of pharmaceuticals and personal care products in the groundwater: a review. Emerg Contam. doi:10.1016/j.emcon.2015.07.001

Sura S, Degenhardt D, Cessna AJ, Larney FJ, Olson AF, McAllister TA (2015) Transport of three veterinary antimicrobials from feedlot pens via simulated rainfall runoff. Sci Total Environ 521–522:191–199. doi:10.1016/j.scitotenv.2015.03.080

Susarla S, Medina VF, McCutcheon SC (2002) Phytoremediation: an ecological solution to organic chemical contamination. Ecol Eng 18:647–658. doi:10.1016/S0925-8574(02)00026-5

Tadeo JL, Sánchez-Brunete C, Albero B, García-Valcárcel AI, Pérez RA (2012) Analysis of emerging organic contaminants in environmental solid samples. Cent Eur J Chem 10:480–520. doi:10.2478/s11532-011-0157-9

Tamminen M, Karkman A, Lõhmus A, Muziasari WI, Takasu H, Wada S, Suzuki S, Virta M (2011) Tetracycline resistance genes persist at aquaculture farms in the absence of selection pressure. Environ Sci Technol 45:386–391. doi:10.1021/es102725n

Tamtam F, Van Oort F, Le Bot B, Dinh T, Mompelat S, Chevreuil M, Lamy I, Thiry M (2011) Assessing the fate of antibiotic contaminants in metal contaminated soils four years after cessation of long-term waste water irrigation. Sci Total Environ 409:540–547. doi:10.1016/j.scitotenv.2010.10.033

Tang X, Lou C, Wang S, Lu Y, Liu M, Hashmi MZ, Liang X, Li Z, Liao Y, Qin W, Fan F, Xu J, Brookes PC (2015) Effects of long-term manure applications on the occurrence of antibiotics and antibiotic resistance genes (ARGs) in paddy soils: evidence from four field experiments in south of China. Soil Biol Biochem 90:179–187. doi:10.1016/j.soilbio.2015.07.027

Tanoue R, Sato Y, Motoyama M, Nakagawa S, Shinohara R, Nomiyama K (2012) Plant uptake of pharmaceutical chemicals detected in recycled organic manure and reclaimed wastewater. J Agric Food Chem 60:10203–10211. doi:10.1021/jf303142t

Tarazona JV, Cesnaitis R, Herranz-Montes FJ, Versonnen B (2013) Identification of chemical hazards for terrestrial plants in the regulatory context: comparison of OECD and ISO guidelines. Chemosphere 93:2578–2584. doi:10.1016/j.chemosphere.2013.09.078

Teeter JS, Meyerhoff DR (2003) Aerobic degradation of tylosin in cattle, chicken, and swine excreta. Environ Res 93:45–51. doi:10.1016/S0013-9351(02)00086-5

Teijon G, Candela L, Tamoh K, Molina-Diaz A, Fernandez-Alba AR (2010) Occurrence of emerging contaminants, priority substances (2008/105/CE) and heavy metals in treated wastewater and groundwater at Depurbaix facility (Barcelona, Spain). Sci Total Environ 408:3584–3595. doi:10.1016/j.scitotenv.2010.04.041

The Commision of the European Communities (2005) Commission regulation (EC) No 378/2005 of 4 March 2005 on detailed rules for the implementation of Regulation (EC) No 1831/2003 of the European Parliament and of the Council as regards the duties and tasks of the Community Reference Laboratory concerning. Brussels

The European Agency for the Evaluation of Medical Products (2000) Environmental impact assessment (EIAs) for veteinary medicinal products (VMPs): phase I. London

Thiele-Bruhn S (2003) Pharmaceutical antibiotic compounds in soils—a review. J Plant Nutr Soil Sci 166:145–167. doi:10.1002/jpln.200390023

Thiele-Bruhn S (2005) Microbial inhibittion by pharmaceutical antibiotics in different soils-dose-responcse relations determinaed with the iron(III) reduction test. Environ Toxicol Chem 24:869. doi:10.1897/04-166R.1

Thiele-Bruhn S, Beck IC (2005) Effects of sulfonamide and tetracycline antibiotics on soil microbial activity and microbial biomass. Chemosphere 59:457–465. doi:10.1016/j.chemosphere.2005.01.023

Thiele-Bruhn S, Peters D (2007) Photodegradation of pharmaceutical antibiotics on slurry and soil surfaces. Landbauforschung Volkenrode 57:13–23

Trapp S (2000) Modelling uptake into roots and subsequent translocation of neutral and ionisable organic compounds. Pest Manag Sci 56:767–778. doi:10.1002/1526-4998(200009)56:9<767::AID-PS198>3.0.CO;2-Q

Trapp S, Mc Farlane JC (1995) Plant contamination: modeling and simulation of organic chemical processes. CRC Press, Boca Raton, FL

Tuhkanen T, Vieno NM, Härkki H, Tuhkanen T, Kronberg L (2007) Occurence of pharmaceuticals in river water and their elimination in a pilot-scale drinking water treatment plant. Environ Sci Technol 41:5077–5084. doi:10.1021/es062720x

Udikovic-Kolic N, Wichmann F, Broderick NA, Handelsman J (2014) Bloom of resident antibiotic-resistant bacteria in soil following manure fertilization. Proc Natl Acad Sci U S A 111:15202–15207. doi:10.1073/pnas.1409836111

US Food and Drug Administration (2013) Guidance for industry #213 new animal drugs and new animal drug combination products administered in or on medicated feed or drinking water of food-producing animals: recommendations for drug sponsors for voluntarily aligning product use conditions with, Federal Register. Rockville

Uslu MÖ, Yediler A, Balcıoğlu IA, Schulte-Hostede S (2007) Analysis and sorption behavior of fluoroquinolones in solid matrices. Water Air Soil Pollut 190:55–63. doi:10.1007/s11270-007-9580-0

Vaclavik E, Halling-Sørensen B, Ingerslev F (2004) Evaluation of manometric respiration tests to assess the effects of veterinary antibiotics in soil. Chemosphere 56:667–676. doi:10.1016/j.chemosphere.2004.02.018

Van Doorslaer X, Dewulf J, Van Langenhove H, Demeestere K (2014) Fluoroquinolone antibiotics: an emerging class of environmental micropollutants. Sci Total Environ 500–501C:250–269. doi:10.1016/j.scitotenv.2014.08.075
Verlicchi P, Zambello E (2014) How efficient are constructed wetlands in removing pharmaceuticals from untreated and treated urban wastewaters? A review. Sci Total Environ 470–471:1281–1306. doi:10.1016/j.scitotenv.2013.10.085
Verlicchi P, Zambello E (2015) Pharmaceuticals and personal care products in untreated and treated sewage sludge: occurrence and environmental risk in the case of application on soil—a critical review. Sci Total Environ 538:750–767. doi:10.1016/j.scitotenv.2015.08.108
Wade Miller G (2006) Integrated concepts in water reuse: managing global water needs. Desalination 187:65–75. doi:10.1016/j.desal.2005.04.068
Wagil M, Białk-Bielińska A, Puckowski A, Wychodnik K, Maszkowska J, Mulkiewicz E, Kumirska J, Stepnowski P, Stolte S (2015) Toxicity of anthelmintic drugs (fenbendazole and flubendazole) to aquatic organisms. Environ Sci Pollut Res 22:2566–2573. doi:10.1007/s11356-014-3497-0
Walsh C (2000) Molecular mechanisms that confer antibacterial drug resistance. Nature 406:775–781. doi:10.1038/35021219
Wang Q, Yates SR (2008) Laboratory study of oxytetracycline degradation kinetics in animal manure and soil. J Agric Food Chem 56:1683–1688. doi:10.1021/jf072927p
Wang J, Lin H, Sun W, Xia Y, Ma J, Fu J, Zhang Z, Wu H, Qian M (2016) Variations in the fate and biological effects of sulfamethoxazole, norfloxacin and doxycycline in different vegetable-soil systems following manure application. J Hazard Mater 304:49–57. doi:10.1016/j.jhazmat.2015.10.038
Warren CR (2013) Quaternary ammonium compounds can be abundant in some soils and are taken up as intact molecules by plants. New Phytol 198:476–485. doi:10.1111/nph.12171
Wei R, Ge F, Zhang L, Hou X, Cao Y, Gong L, Chen M, Wang R, Bao E (2016) Occurrence of 13 veterinary drugs in animal manure-amended soils in Eastern China. Chemosphere 144:2377–2383. doi:10.1016/j.chemosphere.2015.10.126
WHO (2000) WHO global principles for the containment of antimicrobial resistance in animals intended for food. World Health, Geneva, Switzerland
Winckler C, Grafe A (2001) Use of veterinary drugs in intensive animal production: evidence for persistence of tetracycline in pig slurry. J Soil Sediment 1:66–70. doi:10.1007/BF02987711
Winckler C, Engels H, Hund-Rinke K, Luckow T, Simon M, Steffens G (2003) Verhalten von Tetrazyklinen und anderen. Veterinär-antibiotika in Wirtschafts-dünger und Boden. Umweltbundesamt, 44/04
Woodward KN (2008) Assessment of user safety, exposure and risk to veterinary medicinal products in the European Union. Regul Toxicol Pharmacol 50:114–128. doi:10.1016/j.yrtph.2007.10.007
World Health Organization (2001) Monitoring antimicrobial usage in food animals for the protection of human health. WHO, Geneva. doi: WHO/CDS/CSR/EPH/2002.11
Wu N, Qiao M, Zhang B, Cheng WD, Zhu YG (2010a) Abundance and diversity of tetracycline resistance genes in soils adjacent to representative swine feedlots in China. Environ Sci Technol 44:6933–6939. doi:10.1021/es1007802
Wu C, Spongberg AL, Witter JD, Fang M, Ames A, Czajkowski KP (2010b) Detection of pharmaceuticals and personal care products in agricultural soils receiving biosolids application. Clean Soil Air Water 38:230–237. doi:10.1002/clen.200900263
Wu X, Wei Y, Zheng J, Zhao X, Zhong W (2011) The behavior of tetracyclines and their degradation products during swine manure composting. Bioresour Technol 102:5924–5931. doi:10.1016/j.biortech.2011.03.007
Wu X, Ernst F, Conkle JL, Gan J (2013) Comparative uptake and translocation of pharmaceutical and personal care products (PPCPs) by common vegetables. Environ Int 60:15–22. doi:10.1016/j.envint.2013.07.015

Wu X, Dodgen LK, Conkle JL, Gan J (2015) Plant uptake of pharmaceutical and personal care products from recycled water and biosolids: a review. Sci Total Environ 536:655–666. doi:10.1016/j.scitotenv.2015.07.129
Yang Q, Zhang J, Zhu K, Zhang H (2009) Influence of oxytetracycline on the structure and activity of microbial community in wheat rhizosphere soil. J Environ Sci 21:954–959. doi:10.1016/S1001-0742(08)62367-0
Zhang T, Zhang XX, Ye L (2011) Plasmid metagenome reveals high levels of antibiotic resistance genes and mobile genetic elements in activated sludge. PLoS One 6:e26041. doi:10.1371/journal.pone.0026041
Zhang CL, Guo XL, Li BY, Wang Y (2012) Biodegradation of ciprofloxacin in soil. J Mol Liq 173:184–186. doi:10.1016/j.molliq.2012.06.016
Zhang W, Huang MH, Qi FF, Sun PZ, Van Ginkel SW (2013a) Effect of trace tetracycline concentrations on the structure of a microbial community and the development of tetracycline resistance genes in sequencing batch reactors. Bioresour Technol 150:9–14. doi:10.1016/j.biortech.2013.09.081
Zhang DQ, Gersberg RM, Hua T, Zhu J, Goyal MK, Ng WJ, Tan SK (2013b) Fate of pharmaceutical compounds in hydroponic mesocosms planted with Scirpus validus. Environ Pollut 181:98–106. doi:10.1016/j.envpol.2013.06.016
Zhang D, Gersberg RM, Ng WJ, Tan SK (2014a) Removal of pharmaceuticals and personal care products in aquatic plant-based systems: a review. Environ Pollut 184:620–639. doi:10.1016/j.envpol.2013.09.009
Zhang Y, Tang H, Zhou Q, Zhu L (2014b) Effect of temperature and metal ions on degradation of oxytetracycline in different matrices. J Environ Prot 5:672–680. doi:10.4236/jep.2014.58068
Zhao L, Dong YH, Wang H (2010) Residues of veterinary antibiotics in manures from feedlot livestock in eight provinces of China. Sci Total Environ 408:1069–1075. doi:10.1016/j.scitotenv.2009.11.014
Zheng W, Zhang L, Zhang K, Wang X, Xue F (2012) Determination of tetracyclines and their epimers in agricultural soil fertilized with swine manure by ultra-high-performance liquid chromatography tandem mass spectrometry. J Integr Agric 11:1189–1198. doi:10.1016/S2095-3119(12)60114-2
Zhou LJ, Ying GG, Liu S, Zhao JL, Chen F, Zhang RQ, Peng FQ, Zhang QQ (2012) Simultaneous determination of human and veterinary antibiotics in various environmental matrices by rapid resolution liquid chromatography-electrospray ionization tandem mass spectrometry. J Chromatogr A 1244:123–138. doi:10.1016/j.chroma.2012.04.076
Zhu YG, Johnson TA, Su JQ, Qiao M, Guo GX, Stedtfeld RD, Hashsham SA, Tiedje JM (2013) Diverse and abundant antibiotic resistance genes in Chinese swine farms. Proc Natl Acad Sci U S A 110:3435–3440. doi:10.1073/pnas.1222743110
Zielezny Y, Groeneweg J, Vereecken H, Tappe W (2006) Impact of sulfadiazine and chlorotetracycline on soil bacterial community structure and respiratory activity. Soil Biol Biochem 38:2372–2380. doi:10.1016/j.soilbio.2006.01.031
Zuo Y, Lin Y (2007) Solvent effects on the silylation-gas chromatography-mass spectrometric determination of natural and synthetic estrogenic steroid hormones. Comment on "Formation of chlorinated estrones via hypochlorous disinfection of wastewater effluent containing estrone" by Hideyuki Nakamura, Ryoko Kuruto-Niwa, Mitsuo Uchida and Yoshiyasu Terao. Chemosphere 69:1175–1176. doi:10.1016/j.chemosphere.2007.03.065
Zuo Y, Zhang K, Deng Y (2006) Occurrence and photochemical degradation of 17-α-ethinylestradiol in Acushnet river estuary. Chemosphere 63:1583–1590. doi:10.1016/j.chemosphere.2005.08.063

Cadmium Bioaccumulation in Aquatic Oligochaetes Using a Biodynamic Model: A Review of Values of Physiological Parameters and Model Validation Using Laboratory and Field Bioaccumulation Data

Leire Méndez-Fernández, Pilar Rodriguez, and Maite Martínez-Madrid

Contents

1 Introduction 150
2 Physiological Parameters: A Review 152
  2.1 Ingestion Rates 152
  2.2 Assimilation Efficiencies 154
  2.3 Biokinetic Parameters in Cd Bioaccumulation by Aquatic Oligochaetes 155
3 Results 156
  3.1 Physiological Parameters 156
  3.2 Application of the Biodynamic Model to Laboratory Cd Spiked-Sediment Bioassays with Tubifex tubifex 158
  3.3 Application of the Biodynamic Model to Field-Collected Aquatic Oligochaetes or those Exposed to Field Sediments in Laboratory Bioassays 161
4 Discussion 164
5 Conclusions 166
6 Summary 167
References 168

Electronic supplementary material: The online version of this chapter (doi:10.1007/398_2017_1) contains supplementary material, which is available to authorized users.

L. Méndez-Fernández (✉) • P. Rodriguez
Department of Zoology and Animal Cell Biology, University of the Basque Country, Box 644, 48080 Bilbao, Spain
e-mail: leire.mendez@ehu.eus; pilar.rodriguez@ehu.eus

M. Martínez-Madrid
Department of Genetics, Physical Anthropology and Animal Physiology, University of the Basque Country, Box 644, 48080 Bilbao, Spain
e-mail: maite.martinez@ehu.eus

P. de Voogt (ed.), *Reviews of Environmental Contamination and Toxicology Volume 243*, Reviews of Environmental Contamination and Toxicology 243,
DOI 10.1007/398_2017_1

1 Introduction

Oligochaete worms have colonized a wide range of aquatic and semi-aquatic environments, from freshwater to brackish and marine habitats. Different species occupy a variety of microhabitats found in sediments, as well as in aquatic vegetation and decomposing organic matter. Most species are detritivorous, selectively feeding on bacteria, algae and mineral particles rich in organic matter (Coler et al. 1968; Harper et al. 1981a, b; Wavre and Brinkhurst 1971), although a few oligochaetes are predatory. Tubificines and lumbriculids are common oligochaete taxa that can dominate freshwater benthic macroinvertebrate communities (Brinkhurst and Jamieson 1971; Verdonschot 2006), and may account for 50–80% of the biomass in the majority of lakes, rivers and reservoirs (Poddubnaya 1980).

Aquatic oligochaetes have been widely used in sediment ecotoxicity and bioaccumulation studies (see review by Rodriguez and Reynoldson 2011), and their usefulness in Environmental Risk Assessment has been highlighted by Chapman (2001) and Egeler and Römbke (2007). Among those most frequently used are the cosmopolitan species *Lumbriculus variegatus* (Lumbriculidae), *Limnodrilus hoffmeisteri* and *Tubifex tubifex* (Tubificinae), although other species have also been utilized, notably *Branchiura sowerbyi* (Rhyacodrilinae) in tropical regions (Lobo and Espindola 2014). These sediment-dwelling species have been included in standardized protocols for laboratory and field ecotoxicology research: USEPA (2000), ASTM (2005) and OECD (2007, 2008). Their entire life cycle occurs in the sediment, therefore the uptake of pollutants can occur via two different routes: through the integument (porewater) and through digestive epithelia (ingested particles and porewater). In the field, *L. variegatus*, *T. tubifex* and *L. hoffmeisteri* are adapted to a wide range of environmental conditions and can tolerate high levels of pollution (e.g. see Table 3.3, in Rodriguez and Reynoldson 2011).

Sediment-dwelling organisms meet their nutritional requirements from the organic fraction of sediment and the microorganisms associated with this material. Nonetheless, sediment is a poor food source thus massive volumes must be processed in order to obtain sufficient nutrients (Lopez and Levinton 1987). Some oligochaete species are known as upward conveyors, since they ingest huge quantities of underlying sediment and egest feces at the sediment-water interface. These worms build galleries in the sediment, and play an important role in the bioturbation of lakes and rivers through their burrowing activity (Matisoff et al. 1999; Mermillod-Blondin et al. 2001, 2005; Nogaro et al. 2009). This has implications with regard to the physical and chemical properties of the sediment (e.g. sediment "pelletization", changes in redox potential, detritus processing, and nutrient recycling). Bioturbation effects also include the transport of pollutants from the sediment and their release back into the water column (Ciutat et al. 2005; Hunting et al. 2012; Karickhoff and Morris 1985).

Physiological parameters related to digestive processes, such as feeding rates and habits, assimilation efficiencies, and selective feeding on certain sediment particle sizes, have been used for modeling the bioaccumulation of xenobiotics

(Leppänen 1995). Most works quantifying egestion/defecation rates and absorption efficiencies in aquatic oligochaetes date back to the 1970s to 90's; however, in the assessment of metal bioaccumulation, physiological parameters have rarely been mentioned. In 1999, Martinez-Madrid et al. included data on egestion rates, growth rates and cocoon biomass in a *T. tubifex* chronic bioassay, at 13 sites with varying degrees of pollution. More recently, different toxicokinetic models have been used to assess metal assimilation efficiencies in *T. tubifex*, both in spiked-sediment (Gillis et al. 2004; Steen Redeker et al. 2004) and water-only exposures in the laboratory (Steen Redeker and Blust 2004), though only for Cd and Zn.

In the last decade, in an attempt to arrive at a more comprehensive understanding of the relative importance of physiological parameters in the study of trace metal bioaccumulation in invertebrates, the use of a biodynamic model (Luoma and Rainbow 2005) has gained importance. The biodynamic concept provides a framework for explaining how and why trace element bioaccumulation differs between metals, species, and environments. This concept includes the application of bioenergetic-based kinetic bioaccumulation models (Reinfelder et al. 1998; Newman and Unger 2003). It is based on the idea that infaunal organisms can accumulate metals from both water and food, thus bioaccumulation is expected to occur when the rate of metal uptake summed across all sources (solution and diet) exceeds the combined rates of elimination via the gut, excretion or other mechanisms (e.g. epidermal lysosomes, elimination of chloragosomes, autotomy; reviewed by Rodriguez and Reynoldson 2011). However, several difficulties arise in the application of the biodynamic model in sediment-ingesting freshwater oligochaetes due to the complexity of estimating some of the physiological parameters. In sediment-dwelling aquatic oligochaetes, the uptake of contaminants occurs via both epidermal and digestive processes, the latter usually prevailing over the epidermal route (Méndez-Fernández et al. 2014). The biodynamic model has succeeded in explaining metal bioaccumulation in the upward conveyor polychaete *Arenicola marina* at a range of sediment exposure concentrations (Casado-Martínez et al. 2009a, b, 2010a, b). More recently, the biodynamic model has been applied in the oligochaete *L. variegatus* exposed to copper ions and copper oxide nanoparticles (Ramskov et al. 2015).

In the present paper, we reviewed published data on some of the physiological parameters related to digestive processes in three aquatic oligochaete species widely used in standardized bioassays (*L. variegatus*, *L. hoffmeisteri* and *T. tubifex*), as a first step in examining the potential of using the food ingestion rates and assimilation efficiencies reported in the bibliography in a biodynamic model. By using these physiological parameters, we aimed (1) to build a biodynamic model to predict Cd bioaccumulation in detritivorous aquatic oligochaete species, (2) to evaluate the predictions of our model with independently measured tissue residue data in *T. tubifex* exposed to Cd spiked-sediment bioassays in the laboratory and (3) to validate the biodynamic model by comparing predicted *vs.* field Cd bioaccumulation data, in order to identify the limits of its application.

2 Physiological Parameters: A Review

The following sections describe the relevant information on digestive physiological parameters (i.e. ingestion rates and assimilation efficiencies) reported in the scientific, peer-reviewed literature for aquatic oligochaete species, in particular *T. tubifex*, *L. hoffmeisteri* and *L. variegatus*. In this review, we selected those publications that also included data on worm biomass and sediment characteristics (i.e., organic content and/or silt-clay percentage). All data originally given in wet weight were expressed on a dry weight basis, by assuming 90% water content (Gillis et al. 2002; Maestre et al. 2009; Méndez-Fernández et al. 2013). Data on metal concentration originally expressed in micromoles of metal were converted to micrograms. In the following sections the term "tubificids" refers to the oligochaete species of the subfamily Tubificinae.

2.1 *Ingestion Rates*

Aquatic oligochaetes are considered to be continuous feeders; therefore it is assumed that egestion/defecation rates are good approximations of ingestion/feeding rates (Cammen 1980; Martinez-Madrid et al. 1999), the former being easier to measure. Differences in ingestion rates between populations can be due to alterations in feeding behavior, which depend on water and sediment characteristics, such as: (1) sediment particle size, i.e., ingestion rates can be influenced by the selection of a particular sediment fraction and the grain-size frequency distribution of the sediment, (2) sediment total organic content (TOC) or variation in the composition of the associated microbial communities, (3) water temperature and dissolved oxygen concentration, and (4) the presence of toxic chemicals in the sediment that can induce avoidance behavior, with worms either starving or lying on the sediment surface in order to minimize exposure to high levels of pollutants via digestive epithelia or through the body wall, via porewater.

The first three of the aforementioned characteristics have been studied by a number of authors. It is well documented that *T. tubifex* feeds selectively on fine-grained particles (< 63 μm) (Kosiorek 1974; Rodriguez et al. 2001; Tevesz et al. 1980), whereas *L. hoffmeisteri* appears to be more eclectic with respect to the sediment particle size ingested, although it can adopt a strict limivorous regime (Juget 1979). In the field, *T. tubifex* prefers fine sediments (Juget 1979) with some organic content (Verdonschot 1981), while *L. hoffmeisteri* can be associated with either sandy (Juget 1979) or muddy sediments (Birtwell and Arthur 1980; Giere and Pfannkuche 1982; Mildward et al. 2001). *L. variegatus* prefers sandy sediments (Chekanovskaya 1962), although preferential feeding on particles <100 μm has also been reported in this species (Lawrence et al. 2000). Ingestion rates and particle selection by deposit feeders are also thought to be related to the organic matter associated with the particles (Cammen 1980); however, taking TOC as a typical surrogate measurement of the nutritional quality of sediments has not been found to be a good predictor of nutrient availability for worms (e. g. *L. variegatus*: Ankley et al. 1994). Data on the quality and density of the microorganism communities present in the sediment is probably more relevant than TOC for interpreting differences in ingestion rates, but this information is rarely reported in the literature. Wavre and Brinkhurst (1971) studied the diet of three aquatic oligochaetes that ingested the heterotrophic aerobic bacteria present in sediments. The authors indicated that bacteria were the primary food source for the worms and that different species were specialized deposit feeders, selecting detritus with a particular microflora composition. Later, in 1972, Brinkhurst et al. showed the importance of the sediment microflora in the feeding biology of single *vs.* complex cultures of tubificine species.

A critical issue that modifies ingestion rates in oligochaetes is autotomy, a mechanism related to asexual reproduction by architomy in *L. variegatus*, and to the elimination of metals in *T. tubifex*. Autotomy impedes sediment ingestion in *L. variegatus,* for a period of 6–7 days following division of the worm (Leppänen and Kukonnen 1998), as well as the elimination of chemicals in the feces, until the new mouth or anus is formed. During this period, the uptake of chemicals is thought to be restricted to the integumentary pathway.

Under experimental conditions water temperature is usually kept constant, but there is some variability in reports describing the relationship between temperature and digestive parameters. Appleby and Brinkhurst (1970) reported that, in both *T. tubifex* and *L. hoffmeisteri,* defecation rates increased with temperature, and the worms gained weight, with a peak occurring at 18 °C. However, linear increments in feeding rates with increasing temperature have been reported within a range of 7–22 °C (McCall and Fisher 1980), and 4–20 °C (White et al. 1987; Leppänen and Kukonnen 1998). Regarding dissolved oxygen levels, whilst normoxic conditions are usually maintained in laboratory experiments (e.g. >2.5 mg/l; ASTM 2005), temporal hypoxic or anaerobic conditions can occur in the field. The percentage survival of *L. hoffmeisteri* at dissolved oxygen concentrations of 0.1–0.2 mg/l was 24% to 30% at 15 °C and 25 °C, respectively (Aston 1973); whereas Reynoldson (1987) observed that *T. tubifex* maintained under anoxic conditions showed an average survival of 60% for up to 16 weeks at 9.7–12 °C. *T. tubifex* and *L. hoffmeisteri* are known for their capacity for anaerobic metabolism (Gnaiger and Staudigl 1987; Schöttler 1978), and defecation rates seem to be independent of dissolved oxygen down to 0.5 mg/l, only decreasing to zero at < 0.3 mg/l (Volpers and Neumann 2005).

The fourth characteristic that we identified as having an effect on ingestion rates was sediment avoidance, an ecologically relevant response in sediment-dwelling organisms, which has been proposed by different authors as an escape response in the presence of polluted sediment (Amiard-Triquet 2009; Weis 2014). Reductions in the burrowing activity of worms due to the presence of pollutants have been quantified (White and Keilty 1988; Keilty et al. 1988; Meller et al. 1998; Bettinetti and Provini 2002; Rodriguez et al. 2006). In bioassays, this behavior is associated with the presence of worms at the sediment surface over long periods, with scarce movements, and a concomitant reduction in both burrowing activity and the production of fecal pellets (Martinez-Madrid et al. 1999; Méndez-Fernández et al. 2013, 2014). Body biomass reduction in worms exposed to highly polluted sediments has been found to be similar to that shown in worms under starving conditions (worms exposed to calcinated sediments) and has been interpreted as the consequence of sediment avoidance behavior (Martinez-Madrid et al. 1999).

We considered the above miscellaneous factors when searching for relevant data on feeding and defecation rates. Thus, the database used for the biodynamic model includes primarily laboratory data run at 18–22 °C under normoxic conditions, which was the case in most of the laboratory studies. However, data reported under different temperature, food, or oxygen concentration regimes was also considered, for comparative purposes. Data for egestion rates were also differentiated, depending on whether the worms were exposed to unpolluted (bioassay controls, reference sites, and non-toxic conditions) or polluted sediments. Another important issue when considering the variety of data was worm biomass, since larger individuals have higher egestion rates. To tackle this problem, egestion rates were standardized to 1 mg (dw) worm biomass, according to the formula $\mathrm{Yst} = (\mathrm{Wst/We})^{b} \times \mathrm{Ye}$, where *Yst* and *Ye* represent the standard and experimentally recorded egestion rates, respectively; *Wst* is the standard worm weight, 1 mg dw; *We* is the weight of the experimental worms; and *b* is the allometric coefficient that scales the physiological rates to body weight, set at 0.771 by Cammen (1980) for deposit feeders, implying that ingestion rates are controlled by their metabolic requirements. *Yst* values were used as weight-specific ingestion rates in the model (see below).

2.2 Assimilation Efficiencies

In aquatic oligochaetes, urinary excretion is usually disregarded (Brinkhurst et al. 1972) and assimilation efficiency (AE) is calculated as the fraction of absorbed products that is incorporated into body tissues (Penry 1998). However, measuring the parameters required for the estimation of assimilation efficiencies from the sediment matrix is problematic, hence assimilation efficiencies in ecotoxicity studies with benthic invertebrates are difficult to calculate, and some AE data reported are, in fact, absorption efficiencies (Penry 1998).

Wang and Fisher (1999) carried out a revision of the most commonly available techniques for measuring assimilation efficiencies. They looked at two main approaches: radiotracer and mass balance. In freshwater oligochaetes a dual-tracer technique has been applied for the determination of assimilation efficiencies of organic pollutants (Klump et al. 1987; Kukkonen and Landrum 1995; Lu et al. 2004). Another approach, consisting of the use of gamma-emitting radioisotopes, has been extensively used in several aquatic invertebrates, with successful application of the biodynamic model, e.g., in the marine polychaetes *Arenicola marina* (Casado-Martínez et al. 2009a, 2010a) and *Nereis diversicolor* (Rainbow et al. 2009), and in the oligochaete *L. variegatus* (Ramskov et al. 2015).

The mass balance method requires the quantification of total ingestion and egestion. This approach has been used by Brinkhurst and Austin (1979), who calculated AE values for *T. tubifex* and *L. hoffmeisteri* ranging from 2.7 to 5.7%, with an overall mean value of 4.1%. However, this method requires adequate measurement of both total ingested sediment and the fraction digested, which is not possible in most cases. Alternatively, the ratio method proposed by Conover (1966) quantifies the ratios between ash-free dry weight in both food and feces to obtain AE, applying the eq. $AE = [(F' - E')/(1 - E') \times F'] \times 100$, where F' is the ash-free to dry weight ratio of the ingested sediment, and E' is the same ratio calculated for feces. The ratio method relies on the assumption that only the organic component of food is significantly affected by the digestive process (Conover 1966). In the present study, data on sediment and fecal organic content originally reported as LOI% (Loss on Ignition) were converted to total organic carbon (TOC %) using the conversion factor from Nelson and Sommers (1996), which is based on the assumption that organic matter contains 58% organic carbon.

2.3 *Biokinetic Parameters in Cd Bioaccumulation by Aquatic Oligochaetes*

At steady state the uptake of a contaminant is balanced by elimination and growth, to give a value known as constant concentration in the consumer (C_{ss}) (Reinfelder et al. 1998). The concentration of a trace element in a consumer due to food ingestion ($C_{ss,f}$) is given by the function: $C_{ss,f} = (AE \times IR \times C_F)/(k_e + g)$, where AE is the assimilation efficiency (%), IR is the weight-specific ingestion rate, C_F is the chemical concentration in the food, k_e is the loss rate constant after uptake from the food (d^{-1}), and g is the growth rate (d^{-1}). Based on the available literature on oligochaetes, one of the main entry routes for Cd is through the ingestion of sediments (Hare et al. 2001; Warren et al. 1998). More recently, in the paper published by Camusso et al. (2012), the importance of the dietary uptake of Cd (sediment ingestion) in the oligochaete *Lumbriculus variegatus* was demonstrated. Additionally, Méndez-Fernández et al. (2014) showed that ingested sediment in *Tubifex tubifex* was a good predictor of toxicity that was explained by Cd bioaccumulation in the worms.

Regarding the loss rate constant in *T. tubifex*, Gillis et al. (2004) described Cd depuration as a two-compartment model, with the first compartment representing the total body concentration of metal that is easily mobilized, and the second compartment representing the total body concentration of metal that is more tightly bound to worm tissues. These compartments have measurable loss rate constants: k_g, the depuration rate from the gut, and k_e, the depuration rate from body tissues ($k_e = 0.81\ d^{-1}$, $k_g = 56.1\ d^{-1}$, transformed from Gillis et al. 2004). Metal dilution due to organism growth has been ignored in the model since the growth rate constant (adult worms $= -0.03$ to $0.03\ d^{-1}$) in our data was much lower than the loss rate constants (see Wang and Fisher 1999).

To test the accuracy of the biodynamic model in *T. tubifex* bioassays, we used published data on metal tissue residues following long-term Cd exposure (28 days) from Gillis et al. (2002) and Méndez-Fernández et al. (2013), in Cd spiked-sediment experiments. The worms in Gillis et al. (2002) came from a population strain of *T. tubifex* from the Canadian Great Lakes, while the worms in Méndez-Fernández et al. (2013) came from a population strain from a mountain stream in northern Spain. The sediment Cd concentration ranged from 75.3–508.0 μg g^{-1} dw in Gillis et al. (2002) and 0.5–161.9 μg g^{-1} dw in Méndez-Fernández et al. (2013). In the spiking procedures, the mixture and equilibration periods differed between the two studies. In the former, 90-min of mixing was followed by a 3-week equilibration period, and the latter comprised 4-h of mixing followed by a 1-week equilibration period. Sediment digestion procedures were also different, 5% HCl digested in Gillis et al. (2002), and following EPA3052 (65% HNO_3 + HF) in Méndez-Fernández et al. (2013).

Additionally, we wanted to explore the accuracy of the biodynamic model for aquatic oligochaetes exposed to field sediments in laboratory bioassays and for field-collected worms. For the application of physiological parameters in the model, we tested different IR values obtained from reported values in the literature, related to non-toxic *vs.* toxic effects. For those cases where toxicity was not reported, when sediment metal concentrations were higher than the Potentially Effective Concentration (PEC, MacDonald et al. 2000) the sediments were considered as toxic, and if lower than the PEC as non-toxic. When exposed to toxic sediments, worms can demonstrate sediment avoidance behavior, resulting in a lower IR value that we have included in the model. Ramskov et al. (2015) found that in *L. variegatus* the AE value remained approximately constant across an increasing Cu exposure, suggesting a relation to a decreasing IR as a behavioral response. Therefore, in our approach for validation of the biodynamic model, only IR values, not AEs, were chosen depending on exposure to polluted (or toxic) *vs.* unpolluted (or non-toxic) sediments. The AE values used in the model were selected depending on the taxa reported, as defined in the bibliographic source, that is, an AE median value of the three oligochaete species was used for taxa reported as "oligochaetes"; the median of *L. hoffmeisteri* and *T. tubifex* for taxa reported as tubificid worms; and for *T. tubifex,* the median value calculated for this species (related to the < 63 μm sediment fraction).

3 Results

3.1 Physiological Parameters

The laboratory-based defecation rates measured in several aquatic oligochaete species were used as an approximation for ingestion rates (IR) (data for unpolluted and polluted conditions are reported in Supplementary Material Tables S1 and S2, respectively). A summary of the descriptive statistics of the reviewed IR values is

Table 1 Ingestion rates (mg mg^{-1} d^{-1}) in *T. tubifex*, *L. hoffmeisteri* and *L. variegatus* measured under unpolluted *vs.* polluted conditions, and at 18–23 °C

	T. tubifex		*L. hoffmeisteri*		*L. variegatus*	
Ingestion rates	Unpolluted	Polluted	Unpolluted	Polluted	Unpolluted	Polluted
Mean	9.4	4.5	29.5	1.3	15.4	3.7
SE	1.6	1.6	4.1	0.4	3.9	1.4
Median	7.8	2.4	24.5	1.7	11.5	2.1
Minimum	2.6	1.8	3.5	0.5	0.6	0.5
Maximum	26.7	10.4	80.0	1.7	38.8	8.2
n	16	5	23	3	11	5

Abbreviations: SE: Standard Error of the mean
Data sources can be found in Tables S1 and S2

Table 2 Ingestion rates (mg mg^{-1} d^{-1}) in *T. tubifex and L. hoffmeisteri* reported in the literature, at different temperatures and under normoxic or hypoxic conditions

	T. tubifex		*L. hoffmeisteri*	
Ingestion rates	12–14 °C Normoxia	10.5 °C Hypoxia	12–14 °C Normoxia	10.5 °C Hypoxia
Mean	14.5	39.3	3.5	18.4
SE	1.9	4.5	0.5	4.6
Median	15.9	40.6	3.5	17.5
Minimum	8.1	25.3	1.7	4.7
Maximum	19.3	51.3	5.0	36.1
n	6	6	6	6

Source data from Volpers and Neumann (2005)
Abbreviations: *SE*: Standard Error of the mean

shown in Table 1 for *T. tubifex*, *L. hoffmeisteri* and *L. variegatus*, at a temperature range of 18–23 °C and under normoxic conditions. In Table 2, IR values are reported for various temperatures and under both normoxic and hypoxic conditions.

Median IR values (mg mg^{-1} d^{-1}) measured in unpolluted conditions and at a temperature range of 18–23 °C were: 7.8 for *T. tubifex*, 24.5 for *L. hoffmeisteri* and 11.5 for *L. variegatus*. In polluted conditions the values were lower: 2.4 (*T. tubifex*), 1.7 (*L. hoffmeisteri*) and 2.1 (*L. variegatus*). In the absence of pollutants, comparisons between the three species showed significant differences (Dunn's test, $p < 0.05$), indicating the importance of using individual values calculated for each species when possible. Conversely, in polluted conditions, these differences were not significant (Dunn's test, $p > 0.05$), and IR was reduced to similar, low levels in all three species. In all cases, the differences in IR measured in unpolluted *vs.* polluted batches were significant (Kruskal-Wallis test, $p < 0.05$). The maximum IR values recorded in unpolluted conditions were similar for *T. tubifex* (26.7 mg mg^{-1} d^{-1}) and *L. variegatus* (38.8 mg mg^{-1} d^{-1}), and were much higher for *L. hoffmeisteri* (80 mg mg^{-1} d^{-1}). In the presence of pollutants, the maximum IR values for all three species were less than 10.5 mg mg^{-1} d^{-1}, with minimum values being well below 2 mg mg^{-1} d^{-1}.

Table 3 Assimilation efficiencies (AE, %) for three oligochaete species, calculated using the Conover or dual-tracer method (see Sect. 2.2 for details on the methods).

	Conover method		Dual-tracer method	
Assimilation Efficiencies	*T. tubifex*	*L. hoffmeisteri*	*L. hoffmeisteri*	*L. variegatus*
Mean	9.3	10.2	9.7	15.4
SE	1.4	1.0	2.1	2.0
Median	8.7	9.5	9.0	13.0
Minimum	3.4	6.4	2.7	10.9
Maximum	19.6	14.6	16.1	25.6
n	12	9	6	8

Abbreviations: SE: Standard Error of the mean
Data sources can be found in Tables S3 and S4

Ingestion rates at temperatures below 15 °C, which more closely mimic water temperatures in the field in temperate regions, were only available from one study (Appleby and Brinkhurst 1970), which estimated IR using the inverted method to obtain the feces (see Table S1). At 12–14 °C, median IR values in *T. tubifex* were 15.9 mg mg^{-1} d^{-1}, but this value was not significantly higher than the result obtained at 18–23 °C (Kruskal-Wallis test, $p > 0.05$). Contrastingly, IR values were found to be significantly lower at 12–14 °C than at 18–23 °C in *L. hoffmeisteri* (3.5 mg mg^{-1} d^{-1} vs. 24.5 mg mg^{-1} d^{-1}) (Kruskal-Wallis test, $p < 0.05$). Ingestion rates measured in worms under hypoxic conditions (Volpers and Neuman 2005), showed higher median values at 10.5 °C than at 12–14 °C for both *T. tubifex* (40.6 mg mg^{-1} d^{-1}) and *L. hoffmeisteri* (17.5 mg mg^{-1} d^{-1}).

Assimilation efficiency (AE) values, calculated using the Conover and dual-tracer methods, are reported in Supplementary Material Tables S3 and S4. Median AE values using the Conover method (Table 3) were 8.7% for *T. tubifex* and 9.5% for *L. hoffmeisteri*. Using the dual-tracer method, median values were 15.9% for *L. hoffmeisteri* and 13.0% for *L. variegatus* (Table 3), while no data were found for *T. tubifex*. Comparisons between species using the same method were not significant, nor were they for *L. hoffmeisteri* using different methods (U Mann-Whitney test, $p > 0.05$), although the higher variability found in *L. variegatus* AE values may have caused misleading results in statistical analyses. Assimilation efficiencies for each species, independently of the method, ranged from 3.4–19.6% for *T. tubifex*, 2.7–16.1% for *L. hoffmeisteri* and 10.9–25.6% for *L. variegatus*.

3.2 Application of the Biodynamic Model to Laboratory Cd Spiked-Sediment Bioassays with Tubifex tubifex

The physiological parameters used to generate the biodynamic model were obtained from the present literature review, and elimination rates from kinetic experiments by Gillis et al. (2004) with *T. tubifex* (see Table 4). Cadmium tissue residues predicted by the biodynamic model (Table 5) were obtained using several

Table 4 Bioaccumulation parameters used to generate biodynamic model predictions of accumulated Cd concentrations in *T. tubifex*, Tubificids, or Oligochaetes

Species	Parameter	Median; Min-Max	Reference
T. tubifex	IR[a]	Unpolluted: 7.8; 2.6–26.7 Polluted: 2.4; 1.8–10.4	Present review
	AE[b]	9.1; 3.6–19.6	Present review
Tubificids	IR[a]	Unpolluted: 10.6; 1.7–80 Polluted: 1.90; 0.5–10.4	Present review
	AE	9.6; 2.7–36.4	Present review
Oligochaetes	IR[a]	Unpolluted: 11.1; 0.6–80 Polluted: 2.00; 0.5–10.4	Present review
	AE	11.3; 2.7–36.4	Present review
T. tubifex	k_g	56.110	Gillis et al., 2004
	k_e	0.812	Gillis et al., 2004

Abbreviations: *IR* Ingestion rate (g g^{-1} d^{-1}); *AE%* Assimilation efficiency; *Kg* (d^{-1}) the depuration rate from the fast compartment; *Ke* (d^{-1}), the depuration rate from the slow compartment, in a two-compartment model

[a]At normoxic conditions and at 12–23 °C

[b]Only for sediment fraction < 63 μm, see Table S3

combinations within the range of AE and IR values reported in Table 4, depending on the exposure conditions reported in each case.

Measured tissue residues in *T. tubifex* using data from Cd-spiked experiments by Gillis et al. (2002), were in good agreement with our predictions when the physiological parameters used were the maximum values reported in the literature for both AE (19.6%) and IR (26.7 g g^{-1} d^{-1}) (Fig. 1), though when median values were used tissue concentrations were underpredicted by a factor of 2.7–10 (see values in Table 5). However, we found that when median AE (9.1%) and IR (7.8 g g^{-1} d^{-1}) values were applied to the biodynamic model the predicted Cd tissue concentrations were very close to the actual concentrations measured in laboratory bioassays by Méndez-Fernández et al. (2013) for sediment exposures <80.8 μg Cd g^{-1} dw, although the results were underpredicted at the highest sediment exposure (Fig. 1, Table 5).

In the Cd chronic bioassay with *T. tubifex* published by Méndez-Fernández et al. (2013), all worms suffered autotomy when exposed to 161.9 μg Cd g^{-1} dw sediment. The worms were observed at the sediment surface, movements were scarce, and neither fecal pellets nor galleries were present, inferring the existence of sediment avoidance behavior. Therefore, we explored the results of the prediction in the biodynamic model in the presence of a possible reduction in IR by *T. tubifex*. The inclusion of a sediment avoidance factor, through a reduction in the theoretical IR in the contaminated sediment (median value: 2.4 g g^{-1} d^{-1}), with the same AE, resulted in a predicted tissue Cd concentration of 59.4 μg Cd g^{-1} dw, a value much lower than the tissue concentration measured at the highest sediment exposure (6789.3 μg Cd g^{-1} dw).

Table 5 Predicted tissue Cd concentrations ($\mu g\ g^{-1}$ dw) from the biodynamic model using the range and median values of the parameters as presented in Table 4, and measured tissue Cd concentration in 28-day chronic sediment bioassays with *T. tubifex*. Sediment Cd concentrations for each of the exposures are also indicated. Superscripts: 1, Gillis et al. 2002; 2, Méndez-Fernández et al. 2013

Sediment Cd Concentration[1]	Tissue Cd concentration[1]			Sediment Cd concentration[2]	Tissue Cd concentration[2]		
	Predicted				Predicted		
	Median	Range	Measured		Median	Range	Measured
75.3	93.9	12.4–692	254	0.46	0.58	0.1–4.25	0.2
152	189	25.0–1395	1068	13.9	17.3	2.3–128	7.8
208	259	34.2–1912	2635	26.9	33.6	4.4–248	19.2
301	375	49.5–2769	3415	51.1	63.8	8.4–470	28.1
412	514	67.8–3792	3617	80.8	101	13.3–743	248
508	634	83.5–4671	4553	161	202	26.6–1488	6789

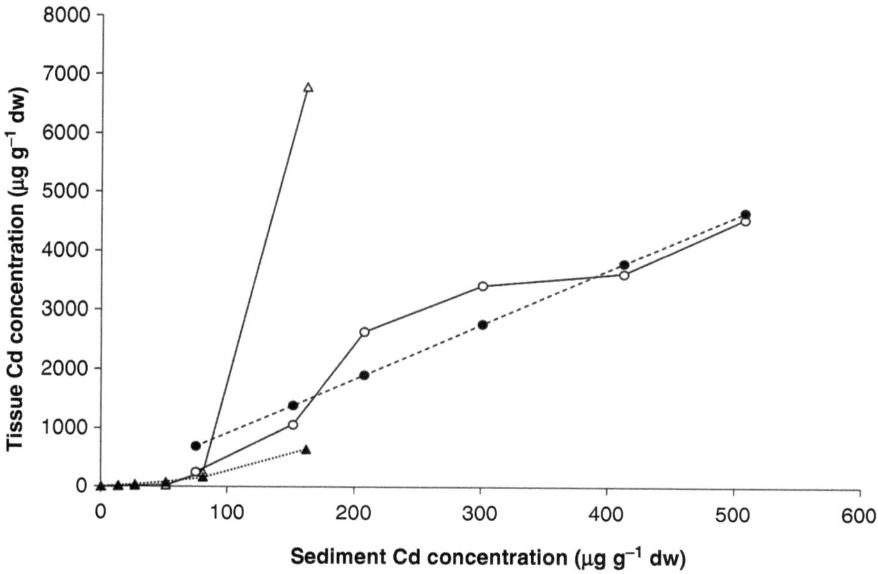

Fig. 1 Tissue Cd concentration measured in *T. tubifex* 28-day sediment exposure and best predicted values from the model plotted against Cd sediment concentration. Symbols: from Méndez-Fernández et al. (2013), measured (*open triangle*) and predicted (*filled triangle*); from Gillis et al. (2002), measured (*open circle*) and predicted (*filled circle*)

3.3 Application of the Biodynamic Model to Field-Collected Aquatic Oligochaetes or those Exposed to Field Sediments in Laboratory Bioassays

Data on sediment Cd concentration and Cd tissue residues reported in papers from several sources have been used to apply the biodynamic model to aquatic oligochaetes, both in laboratory bioassays (Gillis et al. 2002; Méndez-Fernández et al. 2013, 2014, 2015) and field-collected worms (Hernández and Egea 1987; Krantzberg 1994; Protano et al. 2014; Say and Giani 1981; Singh et al. 2007). Reviewed sediment Cd concentrations ranged from 0.03 to 508 µg g^{-1} dw, while measured tissue Cd residues ranged from 0.01 to 6789 µg g^{-1} dw. A significant linear regression was obtained between sediment Cd concentration and measured Cd tissue concentration in aquatic oligochaetes using log-transformed data ($n = 77$, $F = 147.01, p < 0.001$), with a high coefficient of determination ($R^2 = 0.67$, Fig. 2).

We used the information on environmental conditions (i.e. temperature, dissolved oxygen, polluted or toxic sediments) reported in the original papers to refine the physiological parameters utilized in this review as much as possible (Table 4). The values used were: a median IR (g g^{-1} d^{-1}, dw) of 7.3 for *T. tubifex*, 10.6 for tubificids (*T. tubifex* and *L. hoffmeisteri* database), and 11.1

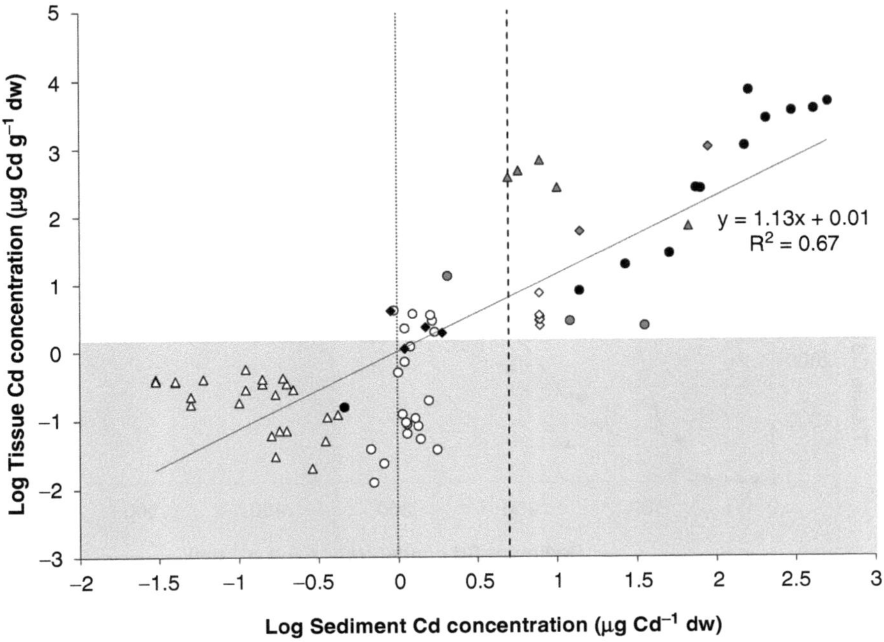

Fig. 2 Measured Cd concentration in freshwater oligochaetes after exposure to field-collected sediments or collected directly from the field. The solid line represents the linear regression between sediment concentration and bioaccumulated Cd. The Sediment Quality Guidelines for Cd proposed by MacDonald et al. (2000) are indicated as the Threshold Effect Concentration value (0.99 µg Cd g^{-1} dw, dotted line) and Probable Effect Concentration value (4.98 µg Cd g^{-1} dw, dashed line); the grey area groups sites where tissue Cd concentration is below the Threshold value for aquatic life proposed by the Oregon Department of Environmental Quality (2007). Symbols: *T. tubifex*: Spiked-sediments from this review (*filled circle*), Méndez-Fernández et al. 2014 (*open circle*), Méndez-Fernández et al. 2015 (*grey filled circle*); *Tubifex* sp.: Singh et al. 2007 (*open diamond*); Tubificids: Hernández and Egea 1987 (*filled diamond*), Say and Giani 1981 (*grey filled diamond*); Oligochaetes: Protano et al. 2014 (*open triangle*), Krantzberg 1994 (*grey filled triangle*)

for undetermined oligochaetes (*T. tubifex*, *L. hoffmeisteri* and *L. variegatus* database) exposed to unpolluted or non-toxic sediments, at temperatures ranging from 12–23 °C. When worms were exposed to polluted or toxic sediment a median IR value of 2.4 was used for *T. tubifex* and 1.90 for tubificids. An AE median value of 9.1% was used for *T. tubifex*, derived from a data series where sediment TOC in the fraction <63 µm was given (see Table S3), while median AEs of 9.6% and 11.3% were used for tubificids and oligochaetes, respectively.

The biodynamic model accurately predicted Cd bioaccumulation across a range of measured tissue concentrations of 0.1–100 µg Cd g^{-1} dw. The results showed that 81.3% of predicted concentrations were within a factor of five of the 1:1 line of the predicted-measured data (Fig. 3). Predicted data were also highly correlated with measured data (Pearson's $r = 0.77$, $p = 0.000$, $n = 49$, log-transformed data)

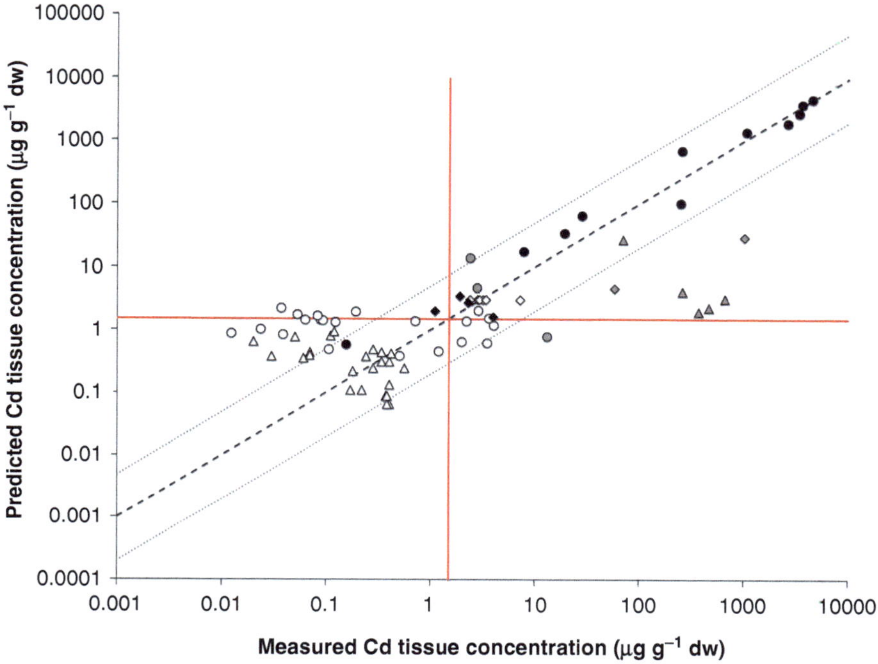

Fig. 3 Relation between predicted and measured Cd bioaccumulation in freshwater oligochaetes exposed to field-collected sediments in bioassays or collected directly from the field. The dashed black line represents a perfect model fit of 1:1 for predicted *vs.* measured Cd accumulation; the dotted grey lines represent error within a factor of five; and the solid red lines indicate the Threshold value for aquatic life proposed by the Oregon Department of Environmental Quality (2007), 1.5 µg Cd g^{-1} dw. Symbols are as in Fig. 2

within a tissue range of 0.1–100 µg Cd g^{-1} dw. However, when considering the entire range of tissue concentrations, the accuracy of predicted-measured values decreased to 67.5% ($n = 77$). Most of the discrepancies between predicted and measured data were associated with the lower range of Cd tissue concentrations, <0.01 µg g^{-1} dw (Fig. 3), with the model overpredicting the results by up to two orders of magnitude for some field sites. At Cd tissue concentrations of >100 µg g^{-1} dw, field data reported from Krantzberg (1994) were underpredicted by a factor of 64–213 (see Fig. 3), although if maximum values of AE and IR are applied, as in Gillis et al. (2002) (see Sect. 3.2), the predicted values are more accurate (only underestimated by a factor of 2.8–9.2). The application of high AE and IR values is supported by the fact that in both of these studies the sediment and the oligochaete population came from the North American Great Lakes (Canada).

4 Discussion

The range of Assimilation Efficiencies (AE) of 2.7–36.4% calculated for three freshwater oligochaetes in the present review, is somewhat in accordance with the range of values (<2.5 to 19%) reported for geophagous earthworms, based on ingested soil organic matter (Curry and Schmidt 2007), and is even closer to the Cd AE reported for deposit-feeder polychaetes, using radiolabelling techniques, e.g. 3.3–43.6% in *Arenicola marina* (Casado-Martinez et al. 2009a). More recently, Ramskov et al. (2015) obtained similar Cu AE values of 30% and 24–41% in *L. variegatus*, after exposure to Cu ions and Cu oxide nanoparticles, respectively. Steen-Redeker et al. (2004) measured Cd AE in *T. tubifex* using radiolabelling techniques, and reported a very low value of 0.09% that the authors attributed to competition with Zn, but this result could also be due to mucus production by the epidermal layer, which acts as a barrier to metals in aquatic oligochaetes (Bouchè et al. 2000; Méndez-Fernández et al. 2014).

The biodynamic model accurately predicted Cd bioaccumulation for *T. tubifex* in laboratory bioassays using Cd-spiked sediments. The best predictions were obtained when using different AE and IR values for each experiment (Fig. 1). Interestingly, for worms exposed to muddy sediments in Great Lakes (Canada), as in Gillis et al. (2002 and 2004: 0.35% OC, 94.4% silt-clay) and Krantzberg (1994), prediction of Cd tissue residues was improved when the maximum values of AE and IR reported in the literature were used in the model, suggesting that the worms may belong to the same source population. This implies that the application of the biodynamic model requires careful and critical selection of physiological parameter values in studies where these figures have not been directly measured. Laboratory measurement of the elimination rate constants for different oligochaete groups may therefore be of interest for improving predictions in relation to field data.

Despite differences in the Cd exposure concentration range utilized in Cd-spiked experiments, similar maximum values for Cd bioaccumulation have been reported, namely 4553.3 μg Cd g^{-1} dw (Gillis et al. 2002) and 6789.3 μg Cd g^{-1} dw (Méndez-Fernández et al. 2013). These values are related to an almost complete inhibition of reproduction in the former study, and to a complete lack of reproduction plus generalized autotomy in the latter. Interlaboratory differences may be due to both population genetics and culture sediment characteristics (e.g. particle size distribution, organic content and natural chemical concentrations, e.g. Reynoldson et al., 1996), and also to differences in detoxification mechanisms (e.g. autotomy and mucous barriers to metal uptake). The use of maximum IR and AE values in the biodynamic model applied to the laboratory data in Méndez-Fernández et al. (2013) improved prediction of the maximum exposure (1488.45 μg Cd g^{-1} dw), but the predicted value was still roughly five times lower than the measured tissue concentration. This possibility has been discarded because a decrease in growth rate was recorded during these experiments, suggesting a stressful situation at the highest concentration. Thus, we interpret that the higher Cd tissue concentration measured could be a consequence of incomplete (or lack of) gut purgation in autotomised worms, or related to difficulties in separating the Cd in tissue from the Cd associated with the mucus barrier. This would result in a higher whole body Cd concentration, but not all of this Cd is metabolically active (Rainbow 2002). The use of lower ingestion rates when there is sediment avoidance, as reported by Méndez-Fernández et al. (2013), did not improve predictions of Cd tissue concentration. This suggests that sediment avoidance prevented Cd bioaccumulation via dietary uptake, but not through the integument, via porewater (which was reported to contain 352.3 μg Cd l^{-1} at the beginning of the experiment). The inclusion of the integumentary uptake route would probably result in better predictions, especially in high pollution scenarios, where there is an excess of metal in the porewater (Méndez-Fernández et al., 2014).

The Sediment Quality Guidelines for Cd proposed by MacDonald et al. (2000) and the Threshold value for aquatic life, 1.5 μg Cd g^{-1} dw, proposed by the Oregon Department of Environmental Quality (DEQ 2007), were used to discuss the environmental risk associated with the predictions of the biodynamic model (Fig. 2). Cadmium concentration in freshwater sediments worldwide usually lies within the range of 0.1–3000 μg Cd g^{-1} dw (Moore and Ramamoorthy 1984), although background concentrations in European stream sediments are normally less than 43.1 μg Cd g^{-1} dw (median value of 0.48 μg Cd g^{-1} dw, $n = 848$; Salminen et al. 2005). In the present review, sediment concentrations ranged from 0.03 to 508 μg Cd g^{-1} dw and exceeded the Cd threshold levels proposed by MacDonald et al. (2000) at 61% of the sites reviewed. These locations largely corresponded to sites where tissue concentration exceeded the threshold value of 1.5 μg Cd g^{-1} dw for aquatic life (Fig. 2). Cadmium sorption and precipitation/dissolution processes are governed by a complex set of environmental variables, such as temperature, oxygen, pH, sediment particle size, and other sediment characteristics (Delmott et al. 2007). These factors could be responsible for the data dispersion in measured Cd tissue residues related to Cd sediment concentration. On the other hand, differences in Cd bioaccumulation can also be due to physiological or genetic variation, as reported for Canadian and Spanish populations of *T. tubifex* in an intercalibration exercise (Reynoldson et al. 1996) and in other studies (Maestre et al. 2009).

Dispersion of the data in the tissue predicted *vs.* measured data representation (Fig. 3), which occurs mainly outside the range 0.1–100 μg Cd g^{-1} dw, could be explained by the variation in analytical techniques used for tissue measurements; by differences in Cd bioavailability (as explained above); or by the presence of metal mixtures. For instance, the presence of Zn reduces Cd uptake in aquatic organisms due to antagonistic effects (Back 1990; Norwood et al. 2007; Steen Redeker et al. 2004). Conversely, some metals may enhance the bioaccumulation of others (Borgmann et al. 2008; De Jonge et al. 2013). Most data overpredicted by the biodynamic model were associated with tissue concentrations below 0.1 μg Cd g^{-1} dw, obtained in bioassays using sediments from areas of the Nalón River Basin (Spain), where the lithology is rich in different metals (mainly As and Hg, but not Cd) (Méndez-Fernández et al. 2015). Nevertheless, predictions were below the threshold value of 1.5 μg Cd g^{-1} dw in all these cases, and do not imply a change in the risk assessment due to Cd bioaccumulation. However, future development of European, national or regional thresholds for metals in both sediment and the tissues of target macroinvertebrate species could alter the risk assessment.

At the other extreme (Fig. 3), high Cd tissue concentrations in field-collected oligochaetes usually resulted in underpredicted bioaccumulation values. In the case of field data from Krantzberg (1994), the existence of summer anoxia in the sediments when the oligochaete worms (Canadian population) were sampled may have caused the deviation. Studies on ingestion rates and assimilation efficiencies under anoxia are contradictory. Volpers and Neumann (2005) suggested that worms stop feeding below 0.3 mg O_2 l^{-1}, however some tubificid species can grow and feed actively despite severe hypoxia (down to 0.005 mg O_2 l^{-1}) and sulfide atmospheres (Fend et al. 2016). Worm behavior in the field under temporal hypoxia or anoxia in the water column is unknown, which makes model predictions difficult. For instance, Fisher and Beeton (1975) demonstrated that oligochaetes burrowed deeper into the sediment during hypoxic conditions and returned to the surface when conditions improved.

Prediction of field bioaccumulation is further complicated by the existence of benthic communities where there are mixed, interacting species. The distribution and abundance of aquatic oligochaete species is determined by the quantity and quality of food available (Moore 1979), and mixed populations of oligochaetes can discriminate between different species of bacteria (Coler et al. 1968). Therefore, the quality of organic matter and the associated microflora in the sediment appears to be more important than the quantity of organic matter, in terms of digestive processes and metal bioaccumulation. Food quality has been shown to have a great effect on metal assimilation in marine bivalves (Wang and Fisher 1996), and the microbial communities in the alimentary tract of deposit feeders are relevant for digestive processes, since intracellular digestion by microbes is an excellent mechanism for achieving differential retention of food components (Lopez and Levinton 1987). Mutualistic interactions between closely related oligochaete species, related to fecal microflora, have been demonstrated in the field (Brinkhurst 1974; Milbrink 1993), and these could also have an effect on assimilation efficiencies.

5 Conclusions

The biodynamic model accurately predicts Cd tissue concentrations in oligochaete worms over a range of measured concentrations, from 0.1 to 100 μg Cd g^{-1} dw, but different ingestion rate and assimilation efficiency values need to be applied under different exposure conditions to improve predictions. The determination of more reliable physiological parameters (including uptake and elimination rates) for use in biodynamic models requires further experimentation, in order to assess the influence of temperature, dissolved oxygen, and probably sediment particle-size distribution. The presence of metal mixtures and other contaminants in field sediments complicates predictions, due to changes in worm behavior (e.g. sediment avoidance) when sediments are toxic and/or to antagonistic or synergic effects between metals. In the current global climate change, an increase in water temperature will play a direct role in the physiology of aquatic organisms, and in the interaction of metals with metabolic processes (Sokolova and Lannig 2008). Consequently, predictions of metal bioaccumulation are likely to be affected. The biodynamic model could be of great help in facilitating prospective risk assessments in these changing scenarios.

6 Summary

This study reviews certain physiological digestive parameters in the literature that could be used to predict tissue residues in aquatic oligochaetes using the biodynamic model. Predictions were evaluated with independently measured Cd bioaccumulation data in sediment bioassays and field oligochaetes. The parameter review focused on three species commonly used in ecotoxicity testing and bioaccumulation studies: *Tubifex tubifex (Tt), Limnodrilus hoffmeisteri (Lh)* and *Lumbriculus variegatus (Lv)*. Median Ingestion rates (g g^{-1} d^{-1}, dw) at unpolluted conditions were 7.8 (*Tt*), 24.5 (*Lh*) and 11.5 (*Lv*), while results were lower (1.7–2.4) at polluted conditions. Assimilation efficiencies ranged from 3.4–19.6% (*Tt*), 2.7–16.1% (*Lh*), and 10.9–25.6% (*Lv*). The biodynamic model accurately predicted Cd tissue concentration in *T. tubifex* exposed to spiked sediments in laboratory bioassays. Comparisons of predicted *vs.* measured Cd tissue concentration in bioassays or field aquatic oligochaetes suggest that the biodynamic model can predict Cd tissue concentration within a factor of five in 81.3% of cases, across a range of measured tissue concentrations from 0.1 to 100 μg Cd g^{-1} dw. Predictions can be refined by using physiological parameter values that have been measured under varying environmental conditions (e.g. temperature, dissolved oxygen). The model can underestimate tissue concentration by up to one order of magnitude when worms are exposed to highly contaminated sediments. Contrarily, predictions overestimate tissue concentration by up to two orders of magnitude when the measured Cd < 0.1 μg g^{-1} dw, although in most cases these predictions do not fail bioaccumulation-based risk assessments, using a tissue threshold value of 1.5 μg Cd g^{-1} dw.

Acknowledgements This investigation has been partially supported by the research project CGL2013-44655-R, sponsored by the Spanish Government, Ministry of Economy and Competitiveness (MINECO). Dr. Leire Méndez-Fernández was supported by a postdoctoral fellowship from the University of the Basque Country. We thank Alexandra Farrell for the English revision. We gratefully acknowledge two anonymous reviewers who helped to improve this manuscript with useful comments and suggestions. Finally, this work was possible thanks to all the "oligochaetologists" who have advanced the understanding of oligochaete biology.

Conflict of Interest The authors declare that they have no conflict of interest.

References

Amiard-Triquet C (2009) Behavioral disturbances: the missing link between sub-organismal and supra-organismal responses to stress? Prospects based on aquatic research. Hum Ecol Risk Assess 15(1):87–110

Ankley GT, Benoit DA, Balough JC, Reynoldson TB, Day KE, Hoke RA (1994) Evaluation of potential confounding factors in sediment toxicity tests with three freshwater benthic invertebrates. Environ Toxicol Chem 13:627–635

Appleby AG, Brinkhurst RO (1970) Defecation rate of three tubificid oligochaetes found in the sediment of Toronto Harbour, Ontario. J Fish Res Bd Canada 27:1971–1982

ASTM (2005) Standard test method for measuring the toxicity of sediments-associated contaminants with freshwater invertebrates. American Society for Testing and Materials—ASTM, Philadelphia, PA. E1706–05

Aston RJ (1973) Field and experimental studies on the effects of a power station effluent on tubificidae (Oligochaeta, Annelida). Hydrobiologia 42:225–242

Back H (1990) Epidermal uptake of Pb, Cd and Zn in tubificid worms. Oecologia 85:226–232

Bettinetti R, Provini A (2002) Toxicity of 4-nonylphenol to *Tubifex tubifex* and *Chironomus riparius* in 28-day whole-sediment tests. Ecotoxicol Environ Saf 53:113–121

Birtwell JK, Arthur DR (1980) The ecology of tubificids in the Thames Estuary with particular reference to *Tubifex costatus* (Claparède). In: Brinkhurst RO, Cook DG (eds) Aquatic oligochaete biology. Plenum Press, New York, pp 331–381

Borgmann U, Norwood WP, Dixon DG (2008) Modelling bioaccumulation and toxicity of metal mixtures. Hum Ecol Risk Assess 14:266–289

Bouchè ML, Habets F, Biagianti-Risbourg S, Vernet G (2000) Toxic effects and bioaccumulation of cadmium in the aquatic oligochaete *Tubifex tubifex*. Ecotoxicol Environ Saf 46:246–251

Brinkhurst RO (1974) Factors mediating interspecific aggregation of tubificid oligochaetes. J Fish Res Bd Can 31:460–462

Brinkhurst RO, Jamieson BGM (1971) The aquatic oligochaeta of the world. Oliver & Boyd, Edinburgh, p 860

Brinkhurst RO, Austin MJ (1979) Assimilation by aquatic Oligochaeta. Int Rev Gesamten Hydrobiol 64:863–868

Brinkhurst RO, Chua KE, Kaushik NK (1972) Interspecific interactions and selective feeding by tubificid oligochaetes. Limnol Oceanogr 17:122–133

Cammen LM (1980) Ingestion rate: an empirical model for aquatic deposit feeders and detritivores. Oecologia 44:303–310

Casado-Martinez MC, Smith BD, DelValls TA, Rainbow PS (2009a) Pathways of trace metal uptake in the lugworm *Arenicola marina*. Aquat Toxicol 92:9–17

Casado-Martinez MC, Smith BD, DelValls TA, Luoma SN, Rainbow PS (2009b) Biodynamic modelling and the prediction of accumulated trace metal concentrations in the polychaete *Arenicola marina*. Environ Pollut 157:2743–2750

Camusso M, Polesello S, Valsecchi S, Vignati DAL (2012) Importance of dietary uptake of trace elements in the benthic deposit-feeding *Lumbriculus variegatus*. Trends Anal Chem 36:103–112

Casado-Martinez MC, Smith BD, Luoma SN, Rainbow PS (2010a) Bioaccumulation of arsenic from water and sediment by a deposit-feeding polychaete *Arenicola marina*: A biodynamic modelling approach. Aquat Toxicol 98:34–43

Casado-Martínez MC, Smith BD, Luoma SN, Rainbow PS (2010b) Metal toxicity in sediment-dwelling polychaete: Threshold body concentrations or overwhelming accumulation rates? Environ Pollut 158:3071–3076

Chapman PM (2001) Utility and relevance of aquatic oligochaetes in Ecological Risk Assessment. Hydrobiologia 463:149–169

Chekanovskaya OV (1962) Aquatic Oligochaeta of the USSRR. Akademiya Nauk SSRR, Moscow, USSR

Ciutat A, Anschutz P, Gerino M, Boudou A (2005) Effects of bioturbation on cadmium transfer and distribution into freshwater sediment. Environ Toxicol Chem 24:1048–1058

Coler RA, Gunner HB, Zuckermann BM (1968) Selective feeding of tubificids on bacteria. Nature 216:1143–1144

Conover RJ (1966) Assimilation of organic matter by zooplankton. Limnol Oceanogr 11:338–345

Curry JP, Schmidt O (2007) The feeding ecology of earthworms—a review. Pedobiologia 50:463–477

Delmotte S, Meysman FJR, Ciutat A, Boudou A, Sauvage S, Gerino M (2007) Cadmium transport in sediments by tubificid bioturbation: an assessment of model complexity. Geochim Cosmochim Acta 71:844–862

De Jonge M, Tipping E, Lofts S, Bervoets L, Blust R (2013) The use of invertebrate body burdens to predict ecological effects of metal mixtures in mining-impacted waters. Aquat Toxicol 142:294–302

DEQ (2007) Guidance for assessing bioaccumulative chemicals of concern in sediment. Oregon Department of Environmental Quality, Cleanup Program, State of Oregon January 31, 2007; updated April 3, 2007

Egeler P, Römbke J (2007) Oligochaeta (microdrile) worms in the environmental risk assessment of pesticides in the European Union. Acta Hydrobiol Sin 31(Suppl):151–162

Fend SV, Liu Y, Steinmann D, Giere O, Barton HA, Luiszer F, Erséus C (2016) *Limnodrilus sulphurensis* n. sp., from a sulfur cave in Colorado, USA, with notes on the morphologically similar *L. profundicola* (Clitellata, Naididae, Tubificinae). Zootaxa 4066:451–468

Fisher JA, Beeton AM (1975) The effect of dissolved oxygen on the burrowing behaviour of *Limnodrilus hoffmeisteri* (Oligochaeta). Hydrobiologia 47:273–290

Gnaiger E, Staudigl J (1987) Aerobic metabolism and physiological responses of aquatic oligochaetes to environmental anoxia- Heat dissipation, oxygen consumption, feeding and defecation. Physiol Zool 60:659–678

Giere O, Pfannkuche O (1982) Biology and ecology of marine Oligochaeta, a review. In: Barnes M (ed) Oceanography and marine biology: an annual review, 20. Aberdeen University Press, Aberdeen, pp 173–308

Gillis PL, Diener LC, Reynoldson TB, Dixon DG (2002) Cadmium induced production of a metallothionein-like protein in *Tubifex tubifex* (Oligochaeta) and *Chironomus riparius* (Diptera): correlation with whole body (reproduction and growth) endpoints of toxicity. Environ Toxicol Chem 21:1836–1844

Gillis PL, Dixon DG, Borgmann U, Reynoldson TB (2004) Uptake and depuration of cadmium, nickel, and lead in laboratory-exposed *Tubifex tubifex* and corresponding changes in the concentration of a metallothionein-like protein. Environ Toxicol Chem 23:76–78

Hare L, Tessier A, Warren L (2001) Cadmium accumulation by invertebrates living at the sediment-water interface. Environ Toxicol Chem 20:880–889

Harper RM, Fry JC, Learner MA (1981a) A bacteriological investigation to elucidate the feeding biology of *Nais variabilis* (Oligochaeta, Naididae). Freshw Biol 11:227–237

Harper RM, Fry JC, Learner MA (1981b) Digestion of bacteria by *Nais variabilis* (Oligochaeta) as established by autoradiography. Oikos 36:211–218

Hernández M, Egea JR (1987) Heavy metal contamination in Guadalix River (in Madrid's industrial belt area). II. Heavy metals in sediments and their bioassimilation by tubificids. Heavy Metal Environ Int Conf 2:172–174

Hunting ER, Whatley MH, van der Geest HG, Mulder C, Kraak MHS, Breure AM, Admiraal W (2012) Invertebrate footprints on detritus processing, bacterial community structure, and spatiotemporal redox profiles. Freshwater Sci 31:724–732

Juget J (1979) La texture granulometrique des sediments et le regime alimentaire des oligochètes limnicoles. Hydrobiologia 65:145–154

Karickhoff SW, Morris SW (1985) Impact of tubificid oligochaetes on pollutant transport in bottom sediments. Environ Sci Technol 19:51–56

Keilty TJ, White DS, Landrum PF (1988) Short-term lethality and sediment avoidance assays with endrin-contaminated sediment and two oligochaetes from Lake Michigan. Arch Environ Contam Toxicol 17:95–102
Klump JV, Krezoski JR, Smith ME, Kaster JL (1987) Dual tracer studies of the assimilation of an organic contaminant from sediments by deposit feeding oligochaetes. Can J Fish Aquat Sci 44:1574–1583
Kosiorek D (1974) Development of *Tubifex tubifex* Müll. in experimental culture. Pol Arch Hydrobiol 21:411–422
Krantzberg G (1994) Spatial and temporal variability in metal bioavailability and toxicity of sediment from Hamilton Harbour, Lake Ontario. Environ Toxicol Chem 13:1685–1698
Kukkonen J, Landrum PF (1995) Measuring assimilation efficiencies for sediment-bound PAH and PCB congeners by benthic organisms. Aquat Toxicol 32:75–92
Lawrence MAM, Davies NA, Edwards PA, Taylor MG, Simkiss K (2000) Can adsorption isotherms predict sediment bioavailability? Chem 41:1091–1100
Leppänen M (1995) The role of feeding behaviour in bioaccumulation of organic chemicals in benthic organisms. Ann Zool Fenn 32:247–255
Leppänen MT, Kukkonen JVK (1998) Factors affecting feeding rate, reproduction and growth of an oligochaete *Lumbriculus variegatus*. Hydrobiologia 377:183–194
Lobo H, Espindola ELG (2014) *Branchiura sowerbyi* Beddard, 1892 (Oligochaeta: Naididae) as a test species in ecotoxicology bioassays: a review. Zoosymposia 9:059–069
Lopez GR, Levinton JS (1987) Ecology of deposit-feeding animals in marine sediments. Q Rev Biol 62:235–260
Lu X, Reible DD, Fleeger JW (2004) Bioavailability and assimilation of sediment-associated benzo[a]pyrene by *Ilyodrilus templetoni* (Oligochaeta). Environ Toxicol Chem 23:57–64
Luoma SN, Rainbow PS (2005) Why is metal bioaccumulation so variable? Biodynamics as a unifying concept. Environ Sci Technol 39:1921–1931
Maestre Z, Martinez-Madrid M, Rodriguez P (2009) Monitoring the sensitivity of the oligochaete *Tubifex tubifex* in laboratory cultures using three toxicants. Ecotoxicol Environ Saf 72:2083–2089
MacDonald DD, Ingersoll CG, Berger TA (2000) Development and evaluation of consensus-based sediment quality guidelines for freshwater ecosystems. Arch Environ Contam Toxicol 39:20–31
Martinez-Madrid M, Rodriguez P, Pérez-Iglesias JI, Navarro E (1999) Sediment toxicity bioassays for assessment of contaminated sites in the Nervión River (Northern Spain). 2. *Tubifex tubifex* reproduction sediment bioassay. Ecotoxicology 8:111–124
McCall PL, Fischer JB (1980) Effects of tubificid oligochaetes on physical and chemical properties of lake Erie sediments. In: Brinkhurst RO, Cook DG (eds) Aquatic oligochaete biology. Plenum Press, New York, pp 253–317
Matisoff G, Wang XS, McCall PL (1999) Biological redistribution of lake sediments by tubificid oligochaetes: *Branchiura sowerbyi* and *Limnodrilus hoffmeisteri/Tubifex tubifex*. J Great Lakes Res 25:205–219
Meller M, Egeler P, Römbke J, Schallnass H, Nagel R, Streit B (1998) Short-term toxicity of lindane, hexachlorobenzene and copper sulfate to tubificid sludgeworms (Oligochaeta) in artifical media. Ecotoxicol Environ Saf 39:10–20
Méndez-Fernández L, Martínez-Madrid M, Rodriguez P (2013) Toxicity and critical body residues of Cd, Cu and Cr in the aquatic oligochaete *Tubifex tubifex* (Müller) based on lethal and sublethal effects. Ecotoxicology 22:1445–1460
Méndez-Fernández L, De Jonge M, Bervoets L (2014) Influences of sediment geochemistry on metal accumulation rates and toxicity in the aquatic oligochaete *Tubifex tubifex*. Aquat Toxicol 157:109–119
Méndez-Fernández L, Rodriguez P, Martínez-Madrid M (2015) Sediment toxicity and bioaccumulation assessment in abandoned Cu and Hg mining areas of the Nalón River basin (Spain). Arch Environ Contam Toxicol 68:107–123

Mermillod-Blondin F, Gérino M, Degrange V, Lensi R, Chassé JL, Rard M, Châtelliers MCD (2001) Testing the functional redundancy of *Limnodrilus* and *Tubifex* (Oligochaeta, Tubificidae) in hyporheic sediments: an experimental study in microcosms. Can J Fish Aquat Sci 58:1747–1759

Mermillod-Blondin F, Nogaro G, Datry T, Malard F, Gibert J (2005) Do tubificid worms influence the fate of organic matter and pollutants in stormwater sediments? Environ Pollut 134:57–69

Milbrink G (1993) Evidence of mutualistic interaction in freshwater oligochaete communities. Oikos 68:317–322

Millward RN, Fleeger JW, Reible DD, Keteles KA, Cuningham BP, Zhang L (2001) Pyrene bioaccumulation, effects of pyrene exposure on particle size selection, and fecal pyrene content I the oligochaete *Limnodrilus hoffmeisteri* (Tubificidae, Oligochaeta). Environ Toxicol Chem 20:1359–1366

Moore JW (1979) Influence of food availability and other factor on the composition, structure and density on a subartic population of benthic invertebrates. Hydrobiologia 62:215–223

Moore JW, Ramamoorthy S (1984) Cadmiun. In: Heavy metals in natural waters. Applied monitoring and impact assessment. Springer series on environmental management. Springer New York, pp. 28–57

Nelson DW, Sommers LE (1996) Total carbon, organic carbon, and organic matter. In: In: Page AL et al. (ed) Methods of soil analysis, Part 2. 2nd edn, Edn Agronomy. 9:961–1010. Am Soc of Agron, Inc Madison, WI

Newman MC, Unger MA (2003) Fundamentals of ecotoxicology, 2nd edn. Lewis, Boca Raton, FL

Nogaro G, Mermillod-Blondin F, Valett MH, François-Carcaillet F, Gaudet JP, Lafont M, Gibert J (2009) Ecosystem engineering at the sediment–water interface: bioturbation and consumer-substrate interaction. Oecologia 161:125–138

Norwood WP, Borgmann U, Dixon DG (2007) Interactive effects of metals in mixtures on bioaccumulation in the amphipod *Hyalella azteca*. Aquat Toxicol 84:255–267

OECD (2007) OECD Test Guideline 225. Sediment–water *Lumbriculus* toxicity test using spiked sediment. Organization for economic coordination and development, OECD, Paris

OECD (2008) OECD Test Guideline 315. Bioaccumulation in sediment-dwelling benthic oligochaetes. Organization for Economic Coordination and Development, Paris

Penry DL (1998) Applications of efficiency measurements in bioaccumulation studies: Definitions, clarifications, and a critique of methods. Environ Toxicol Chem 17:1633–1639

Poddubnaya TL (1980) Characteristics of the life cycle of Tubificidae and Naididae. In: Kothekar VS (ed) Aquatic Oligochaeta worms. Taxonomy, ecology and faunistic studies in the USSR. Amerind Publ Co, New Delhi, pp 97–104

Protano C, Zinnà L, Giampaoli S, Romano-Spica V, Chiavarini S, Vitali M (2014) Heavy metal pollution and potential ecological risks in rivers: A case study from Southern Italy. Bull Environ Contam Toxicol 92:75–80

Rainbow PS (2002) Trace metal concentrations in aquatic invertebrates: why and so what? Environ Pollut 120:497–507

Rainbow PS, Smith BD, Luoma SN (2009) Biodyamic modelling and the prediction of Ag,Cd and Zn accumulation from solution and sediment by the polychaete *Nereis diversicolor*. Mar Ecol Prog Ser 390:145–155

Ramskov T, Amalie T, Marie-Noële C, Henriette S (2015) Biodynamics of copper oxide nanoparticles and copper ions in an oligochaete – Part I: relative importance of water and sediment as exposure routes. Aquat Toxicol 164:81–91

Reinfelder JR, Fisher NS, Luoma SN, Nichols JW, Wang WX (1998) Trace element trophic transfer in aquatic organisms: a critique of the kinetic model approach. Sci Total Environ 219:117–135

Reynoldson TB (1987) Interactions between sediment contaminants and benthic organisms. Hydrobiologia 149:53–66

Reynoldson TB, Rodriguez P, Martinez-Madrid M (1996) A comparison of reproduction, growth and acute toxicity in two populations of *Tubifex tubifex* (Müller, 1774) from the North American Great Lakes and Northern Spain. Hydrobiologia 334:199–206

Rodriguez P, Reynoldson TB (2011) The pollution biology of aquatic oligochaetes. Springer, Dordrecht

Rodriguez P, Martinez-Madrid M, Arrate JA, Navarro E (2001) Selective feeding by the aquatic oligochaete *Tubifex tubifex* (Tubificidae, Clitellata). Hydrobiologia 463:133–140

Rodriguez P, Arrate J, Martinez-Madrid M, Reynoldson TB, Schumacher V, Viguri J (2006) Toxicity of Santander Bay sediments to the euryhaline freshwater oligochaete *Limnodrilus hoffmeisteri*. Hydrobiologia 564:157–169

Salminen R (Chief-editor), Batista MJ, Bidovec M, Demetriades A, De Vivo B, De Vos W (2005). Geochemical Atlas of Europe. Part 1: background information, methodology and maps. Geological Survey of Finland, Espoo

Say PJ, Giani N (1981) The Riou Mort, a tributary to the river Lot polluted by heavy metals. II. Accumulation of zinc by oligochaetes and chironomids. Acta Oecol 2:339–355

Schöttler U (1978) The influence of anaerobiosis on the levels of adenosine nucleotides and some glycolotic metabolites in *Tubifex* sp. (Annelida, Oligochaeta). Comp Biochem Physiol B Biochem Mol Biol 61:29–32

Singh RK, Chavan SL, Sapkale PH (2007) Heavy metal concentrations in water, sediments and body tissues of red worm (*Tubifex* spp.) collected from natural habitats in Mumbai, India. Environ Monit Assess 129:471–481

Sokolova IM, Lannig G (2008) Interactive effects of metal pollution and temperature on metabolism in aquatic ectotherms: implication of global climate change. Clim Res 37:181–201

Steen Redeker E, Blust R (2004) Accumulation and toxicity of cadmium in the aquatic oligochaete *Tubifex tubifex*: a kinetic modelling approach. Environ Sci Technol 38:537–543

Steen Redeker E, Bervoets L, Blust R (2004) Dynamic model for the accumulation of cadmium and zinc from water and sediment by the aquatic oligochaete *Tubifex tubifex*. Environ Sci Technol 38:6193–6200

Tevesz MJS, Soster FM, McCall PL (1980) The effects of size-selective feeding by oligochaetes on the physical properties of river sediments. J Sediment Petrol 50:561–568

USEPA (2000) Methods for measuring the toxicity and bioaccumulation of sediment-associated contaminants with freshwater invertebrates, 2nd edn. The United States Environmental Protection Agency, EPA 600/R-99/064

Verdonschot PFM (1981) Some notes on the ecology of aquatic oligochaetes in the Delta Region of the Netherlands. Arch Hydrobiol 92:53–70

Verdonschot PFM (2006) Beyond masses and blooms: the indicative value of oligochaetes. Hydrobiologia 564:127–142

Volpers M, Neumann D (2005) Tolerance of two tubificid species (*Tubifex tubifex* and *Limnodrilus hoffmeisteri*) to hypoxic and sulfidic conditions in novel, long-term experiments. Arch Hydrobiol 164:13–38

Wang WX, Fisher NS (1996) Assimilation of trace elements and carbon by the mussel *Mytilus edulis*: effects of food composition. Limnol Oceanogr 147:197–207

Wang WX, Fisher NS (1999) Assimilation efficiencies of chemical contaminants in aquatic invertebrates: a synthesis. Environ Toxicol Chem 18:2034–2045

Warren LA, Tessier A, Hare L (1998) Modelling cadmium accumulation by benthic invertebrates in situ: The relative contributions of sediment and overlying water reservoirs to organism cadmium concentrations. Limnol Oceanogr 43:1442–1454

Wavre H, Brinkhurst RO (1971) Interactions between some tubificid oligochaete and bacteria found in the sediments of Toronto Harbour, Ontario. J Fish Res Bd Can 28:335–341

Weis JS (2014) Physiological, developmental and behavioral effects of marine pollution. Springer, Dordrecht

White DS, Keilty TJ (1988) Burrowing avoidance assays of contaminated Detroit River sediments, using the freshwater oligochaete *Stylodrilus heringianus* (Lumbriculidae). Arch Environ Contam Toxicol 17:673–681

White DS, Klahr PC, Robbins JA (1987) Effects of temperature and density on sediment reworking by *Stylodrilus heringianus* (Oligochaeta: Lumbriculidae). J Great Lakes Res 13:147–156

Index

A
Abiotic degradation, 101–103, 129
Abiotic stresses, 2, 17–18
Absorption, 46, 94, 151, 154
Accumulation, 3, 7–10, 12, 14, 27–46, 90, 117, 121, 123, 129
Adsorption, 28, 33–34, 56, 59, 67, 82–83, 107, 109, 113
AE. *See* Assimilation efficiencies (AE)
Aggregation, 55–56, 61–66, 68–69, 72, 76, 78, 80
Agricultural environment, 89–130
Aluminium amelioration, 16
Aluminium sensitivity, 8–13
Aluminium tolerance, 5–9
Aluminium toxicity, 1–18
Antibiotic effects, 115–116, 128
Aquatic environment, 53–83
Aquatic oligochaetes, 149–167
Arsenic, 27–46
Assimilation efficiencies (AE), 150–152, 154–156, 158–159, 162–164, 166–167

B
Bacterial resistance, 91
Bioaccumulation, 149–167
Bioassays, 151, 153–154, 156, 158–165, 167
Biodynamic model, 149–167
Biotic degradation, 90, 101, 129
Boron, 15–16

C
Cadmium, 10, 14, 18, 28, 149–167
Calcium, 10–12, 14, 16, 18
Cancer, 36, 46
Cell wall(s), 4–6, 8–9, 11–13, 16–17, 79
Chronic exposure, 28, 46, 160
Co-morbidities, 28
Concentration levels, 96, 112, 125, 128
Contamination, 30, 32–34, 36–37, 44–46, 55, 107, 109–111, 114, 125

D
Desorption, 28, 32, 34, 45, 67
Deterministic modelling, 70
Diabetes, 46
Dietary uptake, 155, 165
Differential equations, 61–62, 78, 80
Dissolution, 31–34, 55–56, 58, 62–63, 66, 76, 78, 80, 165
DLVO theory, 56, 61, 63–65, 76, 81–82
Drought stress, 2, 8–9, 17–18

E
Elimination rates, 94, 158, 164, 166
Emerging contaminants, 90, 127
Emission estimation, 72
Engineered nanoparticles, 53–83
Environment, 1–18, 27–29, 31–35, 45–46, 53–83, 89–130, 150–151

P. de Voogt (ed.), *Reviews of Environmental Contamination and Toxicology Volume 243*, Reviews of Environmental Contamination and Toxicology 243,
DOI 10.1007/978-3-319-58724-0

Environmental compartment, 28, 56–57, 71–72, 74, 92, 112
Environmental fate, 91, 101
Environmental Risk Assessment, 127, 150

F
Field data, 78, 163–164, 166
Filtration theory, 56, 66–67
Food chain, 27–46, 90, 126
Freshwater organisms, 150–151, 154, 164
Fruit(s), 30, 39, 121

G
GIS-based analysis, 57, 74–75
Groundwater, 28–29, 31–34, 36–37, 45, 66–67, 76, 78, 82–83, 90, 107–113, 118, 120, 128

H
Health, 17, 29–30, 36, 42, 45–46, 54, 90, 101, 116, 126–129
Heavy metals, 2, 10, 14, 17, 28–30, 33–34, 36

I
Ingestion rates, 151–159, 165–167
Iron, 4, 13–14, 31, 34, 54–55, 64, 67, 116
Irrigation water, 28–29, 37, 45

L
Leaching, 12, 16–17, 32, 69, 90–91, 96, 103, 107–113, 118, 128–130
Lead(s), 12, 14, 18, 29–30, 32, 35–37, 42, 46, 54, 62, 64, 67–68, 71, 77, 79, 109, 111, 115–116, 121–122, 128
Leafy, 30, 36–37, 40, 45
Legislation, 126–127
Limnodrilus hoffmeisteri (Lh), 150, 167
Lumbriculus variegatus (Lv), 150, 155, 167

M
Magnesium, 4, 12–14
Manure, 91, 94–108, 110, 112, 115–120, 122, 124–129
Mass flow analysis, 57, 72–74, 76
Metabolites, 90, 94–95, 111, 123
Microbial activity, 33–34, 45, 115–116
Mobility, 10, 27–46, 67, 95, 111, 113
Modelling techniques, 55–58, 60, 66, 68, 76, 79–80
Model predictions, 159, 166–167
Multimedia models, 56, 59, 71–72, 75–76, 80

N
Nitrogen, 4, 11, 14, 18, 115, 119, 121
Nutrient deficiency, 10

O
Occurrence, 36, 75, 95–101, 110–113, 125
Organic content, 75, 122, 129, 152, 155, 164
Oxidative stress, 9–10, 15, 17–18, 123
Oxygen, 28, 31–33, 58, 78, 102, 152–154, 161, 165–167

P
Particle dynamics, 57, 65–66, 76
Pharmaceutical residues, 90, 126
Phosphorus, 13, 15–16, 115, 124
Physiological parameters, 150–158, 161, 164, 166–167
Plant(s), 1–18, 28–29, 32, 34–39, 45–46, 56, 69–70, 73, 76–77, 90, 94, 120–126, 128–129
Plant stresses, 2, 6, 9, 11–12, 16–18, 123
Plant toxicity, 1–18, 28, 36, 45, 90, 124–125
Plant uptake, 4, 6–7, 9–13, 16–18, 35, 45–46, 90, 120–123, 125, 128–129
Pollution, 35, 61, 76, 79, 119, 150–151, 165
Population balance, 56–57, 60–61, 65, 68, 76, 80–81
Probabilistic modelling, 58, 70, 79–80
Processed food, 44–46

R
Reductive, 31–34
Rice, 3–4, 6, 13, 16, 36, 39–46, 122–123
Root exudates, 6–7
Root inhibition, 3, 5, 9, 11, 14, 16–18

S
Sediment(s), 28–29, 31, 33–34, 68, 100–103, 106, 109, 120, 150–156, 159–167
Sedimentation, 58, 61–62, 68–69, 71
Selective feeding, 150
Silicon, 13–16
Simultaneous stresses, 9

Smelter, 30, 45–46
Soil(s), 2, 4–5, 7–18, 28–31, 33–37, 39, 45–46, 55–56, 59–60, 64, 66–67, 71–72, 75, 82–83, 90–91, 94–130, 164
Soil acidity, 4, 10, 14
Soil bacteria, 114–117, 119
Soil compaction, 14, 17–18
Sorption, 35, 69, 90–91, 96, 100–101, 107–113, 116, 122, 125–126, 128–129, 165
Spiked-sediments, 151, 156, 158–162, 164, 167
Stem(s), 30, 39–40, 44, 46, 122

T
Technical compartment, 56, 66–71, 76
Temperature, 7, 32, 44, 58, 103–104, 108, 110, 112, 128, 152–154, 157–158, 161–162, 165–167
Tissue concentration, 124, 159, 161–167
Toxicity, 1–18, 28, 36, 45, 56, 62, 66, 68, 79, 90–91, 95, 114–116, 124–125, 129, 155–156
Transfer coefficient, 71–73
Transmission, 27–46, 126
Transport and fate modelling, 55–83
Transport in soils, 2, 4, 7, 10, 13, 17, 30, 55–60, 64, 66–67, 71, 82–83, 91, 108–110, 112–113, 118, 122
Tuberous, 36–38, 45
Tubifex tubifex (Tt), 150–153, 155–162, 164–165, 167

U
Uptake, 4, 6–7, 9–14, 16–18, 30, 34–35, 37, 45–46, 90, 96, 120–125, 128–130, 150–151, 153, 155, 164–166

V
Veterinary pharmaceutical, 89–130

W
Wastewater, 32, 55, 69–73, 76–77, 90, 95, 100, 120–121, 125, 127, 129

MIX
Papier aus verantwortungsvollen Quellen
Paper from responsible sources
FSC® C105338

If you have any concerns about our products,
you can contact us on
ProductSafety@springernature.com

In case Publisher is established outside the EU,
the EU authorized representative is:
Springer Nature Customer Service Center GmbH
Europaplatz 3, 69115 Heidelberg, Germany

Printed by Libri Plureos GmbH
in Hamburg, Germany